TECHNOLOGY ON THE FRONTIER

Technology on the Frontier

MINING IN OLD ONTARIO

Dianne Newell

University of British Columbia Press
Vancouver 1986

Technology on the Frontier: Mining in Old Ontario

This book has been published with the help of a grant from the Social Science Federation of Canada, using funds provided by the Social Sciences and Humanities Research Council of Canada.

Canadian Cataloguing in Publication Data

Newell, Dianne Charlotte Elizabeth, 1943-
Technology on the frontier

Based on the author's thesis (Ph.D. —
University of Western Ontario): Technological
change in a new and developing country.
Includes index.
Bibliography: p.
ISBN 0-7748-0240-5

1. Mineral industries — Ontario —
Technological innovations — History.
2. Mineral industries — Ontario — History.
3. Ontario — Industries — History.
I. Title.
TN27.05N48 1985 338.26 C85-091441-8

The photographs reproduced in this book are from the following sources: Ralph Greenhill, plates 1, 15; D. G. Hogarth, plate 7; Public Archives of Canada, plates 6, 10, 11; Public Archives of Ontario, plates 2, 4, 5, 8, 9, 13, 14, 16

International Standard Book Number 0-7748-0240-5

Printed and bound in Canada by John Deyell Company

for my son,
Alex Macdougall

CONTENTS

ILLUSTRATIONS

PLATES

PAGE NUMBER

FIGURES

MAPS

TABLES

PREFACE

This book tells about a frontier region in economic transition. Its focus is the successful adoption of new technology to the particular economic and engineering circumstances associated with the newness or frontier nature of Ontario mining to 1890. The subject of mining first attracted my attention as a child growing up in the Ottawa Valley and seeing the evidence of the early mining ventures etched into the local landscape. What started as a childhood fascination has since become a scholarly pursuit.

This book began as a doctoral dissertation. I owe a particular debt to Richard Alcorn, who pointed me in this particular direction, and to Morris Zaslow, my graduate supervisor. Donald G. Paterson did the most of anyone to help me convert the dissertation into the book. All in all, this study required investigation of an array of disparate sources and frequent consultation with experts on various theoretical or technical aspects of the history of technology and mining. I am particularly grateful to the staffs of a number of libraries and archives, most notably the D.B. Weldon Library, University of Western Ontario, Public Archives of Canada, Geological Survey of Canada Library, and the Canada Patent Office Library. Robert M. Vogel, George Richardson, and Norman R. Ball were kind enough to review early drafts of particular chapters. Michael Wayman commented extensively on the metallurgical aspects, and Logan Hovis brought to my attention sources on mining technology that I would have been sorry to miss. In acknowledging all this assistance I in no way wish to imply that those mentioned are responsible for any errors or shortcomings there may be in this final product. On a more personal note, I would like to thank David H. Flaherty, Jane Fredeman, Donald G. Paterson, and Arthur J. Ray for supporting my work at the times when it counted the most.

1

THE PROBLEM

INTRODUCTION

The study of industrial technology and its role in economic and social change represents an important recent trend in Western scholarship, and people interested in Canadian history have begun to discover the value of this subject for understanding Canadian development. While Canada has been a wholesale borrower of technology, Canadians have been simultaneously highly innovative in adapting and combining traditional practices, new ideas, and new machinery to suit specific needs and conditions. The observations of William Marr and Donald Paterson about local adaptation in the case of Canadian transportation are interesting to consider: "So important was the local adaptation that it is difficult to distinguish between a change in technology, say in the railways, and the innovations required for local use, such as the design of gravel beds for track over marsh in northern Ontario, for example."[1]

Nevertheless, few Canadian studies have actually attempted to investigate the specific incidences of local adaptation. To show how specific technological applications and marginal changes made locally affected a young region typically dependent on extractive industries, in this work I focus on Ontario mining operations in the years between 1840 and 1890.

Before 1840, mining was limited to a few basic minerals – coal, copper, lead, and iron ore, for example – and confined mainly to regions in Great Britain and elsewhere in Europe. Some mining occurred in the settled regions of eastern North America, but it comprised mainly non-metallic minerals and was conducted on a small-scale, personal basis for local markets. Over the next fifty years, the mining industry changed dramatically in response to the intellectual ferment in the science of geology and metallurgy and the rapidly increasing demands of the industrial age for more and different minerals. In particular, the growth of railways, iron shipbuilding, metal-making, and metal-working had greatly escalated the demand for iron ore, copper, lead, and complex ores. Consequently, there was a rapid transformation of mining technology from hand methods to mass-mining techniques.

Within a few decades, world demand and supply in minerals increased greatly. For example, copper sources before 1840 were limited to Sweden, Cornwall, and Devon. With the opening of new copper regions in Chile, Spain, Australia, and Upper Michigan, world production more than doubled by 1865 and had increased by one-hundredfold by the century's end.[2] Likewise, the world's average annual production of gold leapt from £5 million in 1850 to £25 million by 1875 as a result of the opening of the American and Australian goldfields. Petroleum had only become commercially valuable in the 1860s, principally as an illuminant, but by 1890 oilfields had been opened throughout the United States, in Eastern Europe, Mexico, Australia, and elsewhere.

The industrial growth and transportation revolution therefore provided both the occasion and the means to make mining on the Ontario frontier economically possible. Everywhere large-scale, high-speed production enterprises were becoming more common. But, for such marginal areas as Ontario, increased output was not solely the result of the efficiencies of large-scale production or of the long-run decline in transportation costs. Of great significance was the ability to open previously higher cost areas to profitable production.[3] This extension of accessible mining frontiers was accomplished by the new technological capacity to work more difficult or lower-grade and more complex deposits, as the case studies will show.

Fifty years after a Geological Survey was appointed to enquire into the economic geology of the newly created Province of Canada, mining was an established industry, and territorial expansion had begun into the "New Ontario" of the north. Confirming this new northern mining era, the Ontario Bureau of Mines and School of Mining were established to put the management of provincial mineral resources on a sound and formal basis. By 1900, Ontario had entered the first rank of world metallic-mineral producers. However, its modern mining operations were largely standardized, concentrated in the northern regional setting, and dominated by United States' technology. Consequently, while they are of undeniable concern to historians of mining, for the specific purposes of this study the mining activities of the post-1890 period are less important for tracing the evolution of mining techniques than are those of the earlier transition period.

Between 1840 and 1890, as mines were brought into production all over North America, an unusually large number of minerals were discovered and worked in Old Ontario in three contrasting physical and social settings by both foreign and domestic owners. New mining operations included silver, copper, nickel, iron ore, hardrock gold, lead, phosphate of lime (apatite), mica, and petroleum and salt extraction. Because of this variation, Ontario was the principal mining area of the country. The variation and diversity of mining is of interest because it entailed the application of a large range of methods and machinery.

In undertaking this particular study I hope to offer a fuller picture of the circumstances under which individuals in a small, open economy like Canada's were induced to import, even to develop, technological solutions in the search for and exploitation of resources in demand on world markets. The study also provides a new perspective on such related issues as shifts in world prices and markets for minerals, the revolution in communications and transportation networks, resource-based community development, and the growth of public policy on resource development, all of which have been, and continue to be, crucial to the opening of new staple export regions to development.

PATTERNS OF TECHNOLOGICAL CHANGE

Scholars of American and British history and related disciplines have understandably devoted much attention to the changes in nineteenth-century technology. Their findings are worth investigating in a Canadian context. There are two major aspects of technological change, the first of which is the temporal. Technological inventions, by which are meant the breakthroughs, and innovations, which are the practical applications of inventions, tend to cluster in time, largely in relation to the patterns of economic growth and factor price trends. The economics of new technology generally determined the timing of its first introduction and the speed of its spread.[4]

Although historians do not always agree on the timing of industrialization as it spread during the eighteenth and nineteenth centuries, there is agreement that the process of industrialization involved transfers of technique from one economy to another even although there were important sources of growth of domestic origin. It is fairly certain that in the United States a major turning point was reached in the 1840s with the discovery of the rich anthracite coal deposits in eastern Pennsylvania.[5] Prior to 1840, the lack of an inexpensive, efficient fuel had been a serious material constraint on the widespread adoption of steam technology and on the spread of factories – both of which are usually considered essential elements of industrialization. Joseph Schmookler and many others have investigated United States patent statistics to determine the trends in American inventive activity during the industrial era.[6] Not surprisingly, the inventions of the early industrial period were based largely on crafts and simple mechanical engineering, whereas by the late nineteenth century there had occurred a marked shift towards the newer industries based on physics and chemical engineering.[7]

The path of technological diffusion and adoption has been neither as smooth nor as automatic as scholars such as Schmookler imply. On the contrary, time-lags in innovation diffusion and adoption have been routine. There were monumental inventions, such as the steam engine, that were truly unique, but there were also the interdependent, mutually supportive, and subsidiary improvements to existing techniques required to bring basic inventions into practical commercial uses. The more revolutionary the new technology, the greater the incidence of modification and improvement, of innovation, a finding which might help to explain the exponential growth in patent applications during the late nineteenth century, when so much new basic technology was being developed.[8] Excellent examples of technological lag are to be found in Nathan Rosenberg's classic study of the American machine tool industry, Carroll Pursell's and Peter Temin's investigations of early stationary steam engines in America, Charles Hyde's work on the British iron industry, Larry Lankton's study of compressed-air drills in Michigan copper-mining, and Jennifer Tann's work on steam technology in the process industries during the industrial revolution in Great Britain.[9]

Of interest to this investigation of technological change in frontier areas is the evidence that innovation also occurred in cases where seemingly outmoded technology persisted in the face of new advances. When Louis Hunter examined waterpower in the age of the steam engine, for example, he discovered that the rise of a large, flourishing industry engaged in the development of a distinctive and distinguished class of waterwheels accompanied the general decline in the use of waterpower.[10] Richard Shallenberg and Bruce Seely reached similar findings when they studied the American charcoal iron industry, an industry which was as aggressive in its pursuit of improvements in blast-furnace technology as were the mineral-fuel users who ultimately replaced it.[11]

Despite the tremendous increase in invention and innovation in the second half of the nineteenth century, only a small proportion of firms were willing or able to be innovators. Engineering efficiency was not always economic efficiency. Those few firms which undertook the major share of research and development costs were usually motivated by a desire to gain immediate, direct competitive advantage over others. These innovative firms became the established leaders, while the others in their industries, who obtained what amounted to a "free ride," were the followers.[12]

Invention and innovation in the nineteenth century did not, however, involve deliberate research and development in modern terms, and workers who sensed a threat to their independence or jobs often resisted the introduction of new technology. But in successful transformations of knowledge, "learning-by-doing" and the "bottom-up" approach played a major role in generating productivity-raising designs for new techniques and for modifying existing ones. The importance of this type of indigenous, localized invention, which has been studied by Paul David, becomes even clearer in examining what Robert Allan terms "collective invention."[13] In this important nineteenth-century process, firms in an industry, such as iron and steel, collectively invent in the course of routine production and then make available to their competitors the results of their new designs and techniques (or, as indicated earlier, cannot hide the technical advantage from them). Overall, there is reason to suspect that perfected human skills and the process of collective invention may have been the most important sources of new technology in the nineteenth century. Certainly there exists convincing evidence that in the United States before 1900, individual rather than deliberately organized corporate effort accounted overwhelmingly for patented inventions.[14] The major swing to corporate research and development as a source of patents did not occur until at least the 1920s.

This individuality points to the second major aspect of technological change: the geographic. Inventions and innovations tend to cluster in geographic locations and diffuse outward from these innovation centres. Historical geographers and rural sociologists have identified this as a relatively common happening, though they often treat technological diffusion as primarily a matter of information dissemination and therefore consider only the externally invented and perfected innovations that were not further improved upon locally during the course of their acceptance. This line of investigation, pioneered by Swedish geographer Torsten Hägerstrand, has become more dynamic in the hands of urban geographers such as Allan Pred.[15] Innovations often centred in urban places. This is where investment capital was available, manufacturing was sizeable, and the repertory of new techniques and imitatible successes was accumulated. The industrial inventions and innovations which were developed and then adopted in specific urban places produced initial advantages which, in turn, sustained the cumulative growth of those places as metropolitan centres and consequently led to additional inventions and innovations.

Typically, the interaction between urbanization and technological invention and innovation is presented as a feedback process contributing to increasing geographical concentration. However, one of the reasons why stationary steam technology, for example, was so important for economic development was because it did not necessarily lead to geographical concentration. Initially, at least, steam provided an essentially dispersive capacity — for instance, in mining and lumbering—because it freed early industrial entrepreneurs from a dependency on sites with waterpower potential.

The question of how a particular regional setting would actually effect the search for new methods and machinery in a specific direction has been a significant one for historians. Over the long run, technological advances in the industrial era have tended to be capital-using and/or natural resource-using, and labour-saving, labour having become the relatively expensive factor. Perhaps the best-known explanation concerning the issue of technological bias in the nineteenth century is the notion of labour scarcity. According to this thesis, the relative labour shortage and high wage rates relative to other input prices, plus an abundance of natural resources, caused the United States to invent and adopt mechanized, labour-saving and capital-intensive techniques more rapidly than Britain. First posited by H.J. Habakkuk, the labour scarcity thesis has been highly influential in guiding the research of economic historians and historians of technology for the past two decades.[16] Recently, Paul David has also demonstrated the importance of land tenure; scale factors; the costs of replacing outmoded physical plants; and changing relative factor prices, especially rising labour prices and falling capital prices, in decisions about adopting labour-saving technology.[17]

Though historians of labour in the twentieth century recognize major and minor earlier occasions on which workers themselves either engaged in technological innovation or intervened to try to prevent the adoption of new machinery, some conclude that a major force for introducing new technology, particularly in the twentieth century, was a bid by management to impose controls on the workforce, even to "de-skill" labour.[18] Wherever the balance lies in these individual cases, even if economic motives were primary, workers were usually key in the successful adoption of new techniques. A single dramatic instance is the introduction of practical compressed-air drills underground in Michigan copper-mining in the early 1880s. The introduction of this technological advance resulted in 20 per cent more copper ore being raised with 20 to 30 per cent fewer workers; however, it had taken ten years to find a particular drill that did not threaten to destroy important social and economic patterns associated with hand-drilling.[19] Ironically, a critical factor in the spread and adoption of such new labour-saving techniques as the compressed-air drill was the migration of workers already trained in its use and capable of passing their skills on to others.[20]

As a final note on the geography of technological change, its role in the regional economic growth of export economies such as Canada should be cited. From the standpoint of those in a new export region, the demand for the export commodity was an exogenous factor, but processing and transfer costs were not. Thus, as a region matured around an export base, such as minerals, there would be an attempt to improve the competitive cost position of the export community by improving the techniques of production and to reduce the costs of output and of transportation.[21] Interestingly, when a number of adopted or adapted technological innovations combined to a single method, its observed success was often imitated by other producers and could become the standard, or the system, for the entire region. The development of regional systems indirectly made an industry more efficient, as has been demonstrated by Richard Shallenberg for the American charcoal iron industry, William Gates for the Michigan copper-mining district, and Harold Innis for the major mining regions of Canada.[22]

TABLE 1
DATES OF TURNING POINTS IN CANADIAN PRICE CYCLES, 1868-1891

Period	*Phase*	*Reference Date*
1	Low	November, 1869
2	High	April, 1874
3	Low	August, 1879
4	High	April, 1882
5	Low	September, 1886
6	High	May, 1891

Source: K.W. Taylor and H. Mitchell, *Statistical Contributions to Canadian Economic History*, vol. 2 (Toronto: Macmillan, 1931), App. table B, p. 93.

In Canada, there were a few years of economic prosperity, but otherwise the period from 1868 to 1891 was one of protracted price declines.

In Canada, the history of technology is still in its infancy, yet already historians, led by W.A. MacIntosh and Harold Innis, have created a powerful image of the general role of technology in Canadian economic and social development.[23] As is well known, the major scientific discoveries and key inventions of the new industrial age occurred primarily in advanced regions and then spread by way of persons, publications, and hardware to be applied as innovations in the less advanced regions. The major mining countries had already reached a high level of industrial development by the second half of the nineteenth century and compared with these regions Canada's industrial and mining development was different. As an open economy, it was a supplier of staple commodities that depended, in some measure, on the more advanced economies for technology, skilled labour, capital, and markets. As indicated below, the Ontario mining districts were especially vulnerable to shifts in mineral markets and to the discovery of new external sources of supply.

For Ontario mineral producers, the price of minerals was the world price, even in the case of petroleum and salt, which always had largely domestic markets. As with most prices, the world price of minerals declined over the second half of the nineteenth century. In Canada, there were a few years of economic prosperity, from 1867 to 1873 and 1880 to 1883, but otherwise the period from 1868 to 1891 was one characterized by O. J. Firestone as being of "highly unsettled economic conditions and protracted price declines."[24]

In addition to the general economic situation, there were other economic factors influencing prices of Canadian minerals. For instance, in the early 1870s Canadian petroleum prices fluctuated sharply owing to the competitive nature of the young oil industry and the great number of newly discovered oilfields coming into production in the United States and elsewhere in the world. From a low of 60 cents a barrel in 1862, petroleum leapt to $8 in 1864, dipped to $4 in

1872, and sank to less than $1 in 1873.[25] By the next decade, the Standard Oil Trust in the United States controlled domestic production and also world

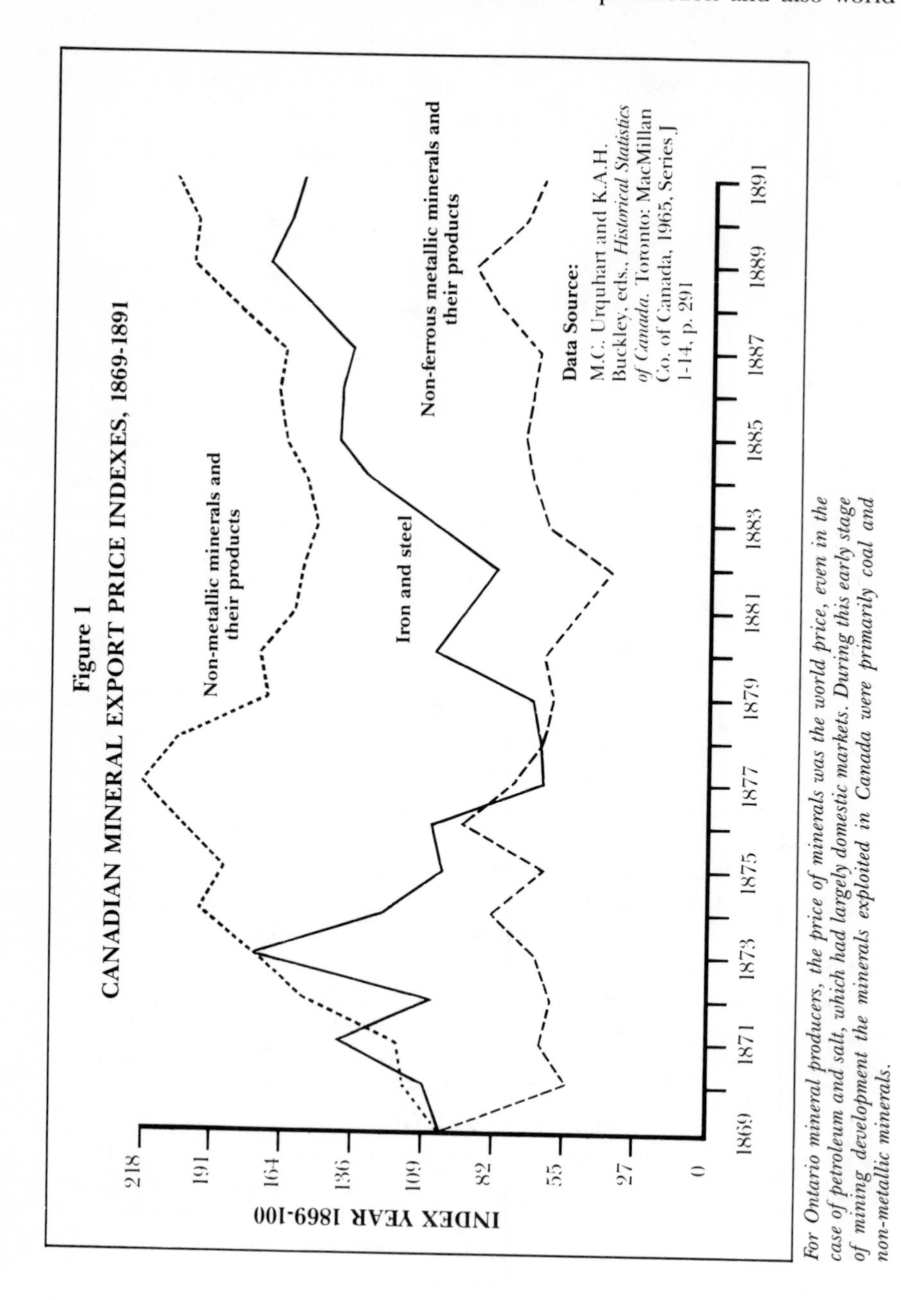

Figure 1
CANADIAN MINERAL EXPORT PRICE INDEXES, 1869-1891

For Ontario mineral producers, the price of minerals was the world price, even in the case of petroleum and salt, which had largely domestic markets. During this early stage of mining development the minerals exploited in Canada were primarily coal and non-metallic minerals.

markets. Not only was Ontario petroleum excluded from the foreign (including United States) markets, but also since the United States was not prohibited from supplying the Canadian market, it competed with Ontario petroleum for that market too.[26] A somewhat similar story existed for copper. Canadian copper prices plunged dramatically in 1869. In that year it lost its American markets for copper concentrates owing to the pricing arrangements and protective tariff imposed on behalf of the powerful Michigan copper producers.[27] Within a few years, the world supply of copper had increased more than demand, and as it did, the price fell away. A major downward price plunge followed in the wake of the worldwide financial crisis of 1883.

The influences of these modern industrial transformations and Canada's vulnerability to outside economic forces have been stressed in Innis's pioneering studies on the opening of Canadian natural-resource frontiers to development.[28] While Innis focused on the critical role of Canadian geography and the nature of individual staples – fish, fur, and minerals – other scholars, from Albert Faucher to Morris Zaslow and Mel Watkins, J.M.S. Careless and A.R.M. Lower, have tested and refined Innis's "staple thesis" to show how geographical shifts in investments, constant influxes of new populations, and shifts of metropolitan centres also affected Canada's vulnerability to outside economic forces.[29]

It was this reliance upon imported technology that inevitably involved Canada in experimentation, modification, and improvement to suit local conditions and solve bottlenecks.[30] Such technological adaptation was, I would argue, a key to the economic success of frontier regions. Local inventions and regional technological systems might be imitated in other regions, once the information became available and assuming the technical conditions were similar. Studies of Canadian invention, such as they are, tend to focus on those major important technological advances which had widespread general application. There exist a number of examinations of imported technology which assume that Canadians automatically adopted the new techniques once they had been introduced. Such is the case with Elwood Moore's study of the contributions of the United States to the Canadian mining industry, for instance.[31] The more common notion, however, is that the history of Canadian technology is one of continual selection and adaptation of imported techniques.[32]

The problem of time-lag in the adoption of technological advances, which is the focus of Richard Pomfret's study of the mechanical reaper and its adoption in Ontario, and the complete failure to adopt new technology, which is the subject of separate studies on the decline of the Canadian wooden shipbuilding industry by Albert Faucher and C.K. Harley, are issues central to understanding the process of technological change in a country like Canada.[33] Pomfret looked at the timing and rate of adoption of mechanical reapers in Ontario in terms of the improvements in the original invention which were required and in terms of farm size. Faucher examined the shipbuilding industry at Quebec by placing it in its larger continental context. Harley focused on the immobility of labour and other factors employed in Atlantic shipbuilding. In their conclusions all three scholars reject the traditional, Schumpterian view that the failure to adopt new techniques (and hence the persistence of old techniques) can be explained away as prejudice or ignorance. And the investigations of mining in Ontario provide corroborative evidence.

THE PROCESS OF TECHNOLOGICAL CHANGE

At this point it is possible to offer a framework of analysis with which to study technological change on the mining frontier of Old Ontario. With technological change, two distinct processes and several stages of development operate.[34] The first of these concerns the method of spread. For technological change to occur, there must exist a general pool of *available* technology from which a new technique derives and receives *publicity* by way of people, publications, and even hardware. The second process involves the rate of spread. At this stage, a potential innovator becomes *aware* of the new technique and may become *interested* in it. To *evaluate* its merits, essentially performing a cost-benefit analysis, he will test it. The *trial* stage might lead to modifications, improvements, or even *local inventions* before the new technique is *adopted.* Local inventions, when they occur, and if successful, might contribute to the general pool of technology. And any innovations successfully introduced at a single operation might be readily imitated by other operations in the region.

The remainder of this study on technological change in mining in Old Ontario is organized around the sequence of developments in the process of technological change. Chapter Two investigates the international pool of appropriate machinery and methods available to provincial mine operators as described and publicized in dozens of contemporary English-language publications on mining, engineering, and technology. Chapter Three focuses on a traditional mode of technological diffusion by examining the migration to, and within, the province of experienced mining men; professional mining engineers, geologists and surveyors; and mining equipment promoters. It then speculates on the potential of mining, engineering, and government publications for publicizing the latest machinery and methods. The next three chapters concentrate on the reception of technological change by Ontario mining operators.

Each case study chapter looks at the course and causes of technological change over the life-cycle of mining operations in one of the then-principal mining districts of the Upper Great Lakes, southeastern, and western peninsula. The unity of theme is in the appeal to the frontier, small-scale nature of Ontario mining, even in the case of petroleum, where Ontario, along with Pennsylvania, was the world frontier.

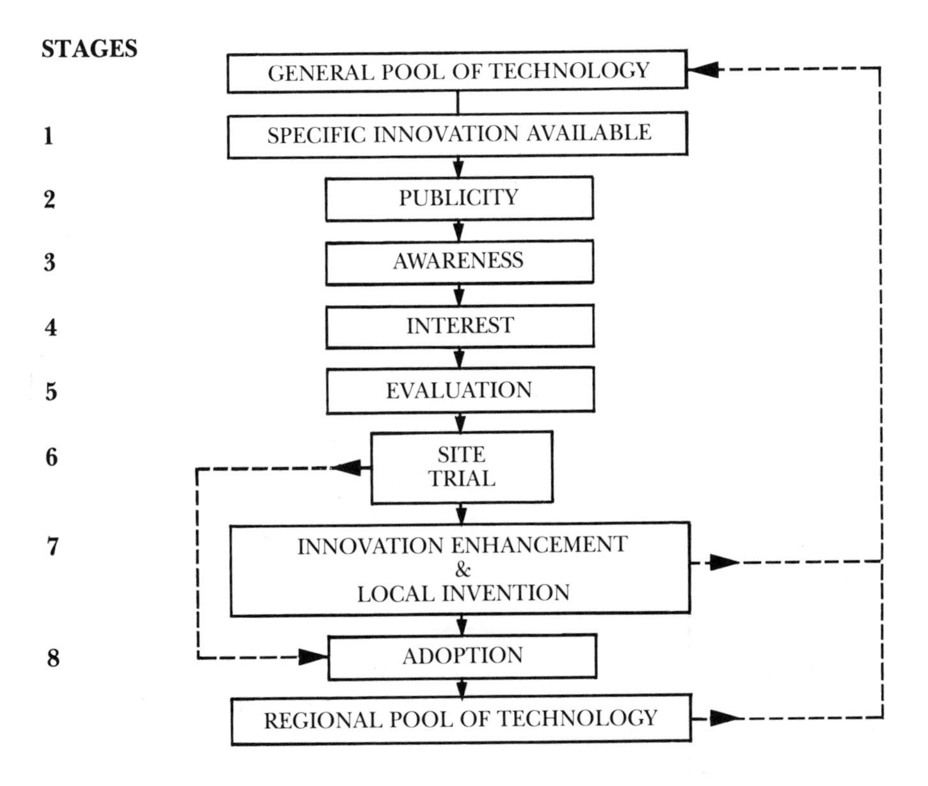

Figure 2. Technological Change as a Sequence of Developments. *The process of technological change is a dynamic one and therefore ought to be studied as a sequence of developments. To be sure, the actual path that diffusion and innovation follows is not usually so clear cut as is presented here. This model is offered simply as a guide for examining technological change occuring within any period or setting.*

The term "mining activity" as used here applies to all the operations likely to take place at the minesite. It includes the extraction of minerals by excavation of solid materials by trenching in the open air, or sinking shafts from which are driven tunnels, and stoping; or drilling and pumping liquids and gas; and it applies also to the operations which continued processing through the prime metal, or ingot, stage; crude oil refinery; and salt works.[35] Because coal resources to provide the energy for processing were lacking in the province, minerals (other than oil and salt) were almost certain to be shipped elsewhere for final processing and use. Common open pit and quarry operations such as extracting building stone, sand, gravel, cement, terra cotta, clay, and plaster are omitted from consideration. These materials bore such low value in proportion to their weight that their cost of transportation restricted their markets to local uses. They were economically important extractive operations in total, but they were so simple in technique, and so obviously tied to local markets, that it would not

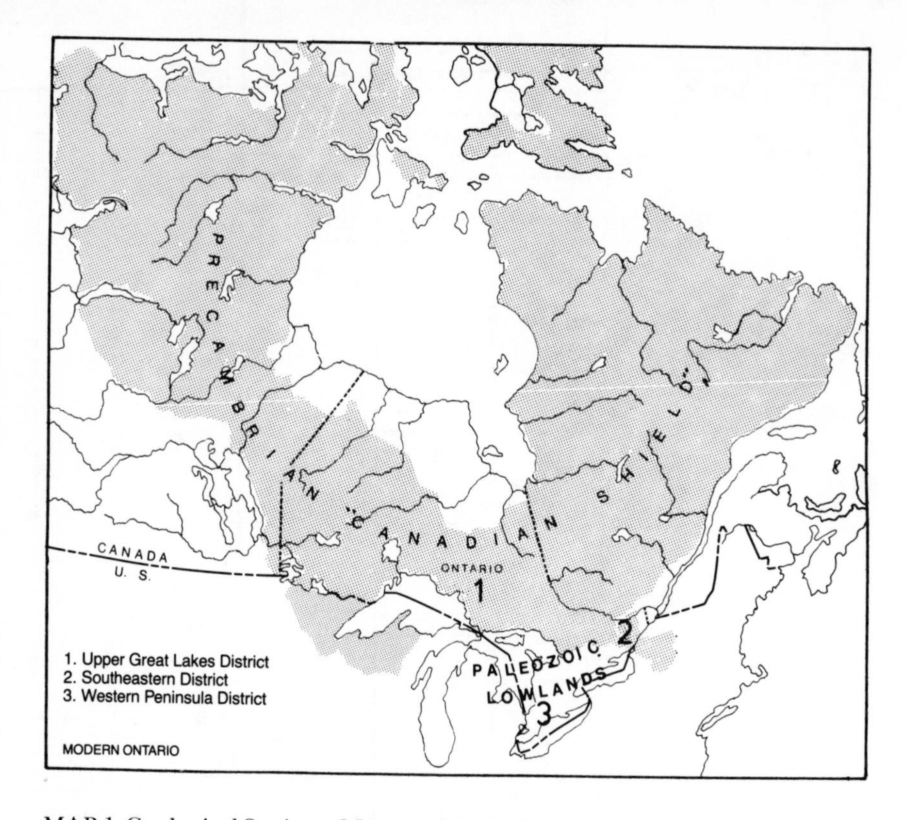

MAP 1. Geological Setting of Ontario Mining Districts, 1841-1891. *Between 1841 and 1891, as mines were being brought into production all over North America, an unusually large number of minerals were discovered and worked in Old Ontario in three contrasting physical and social settings.*

be feasible to include them in a detailed study of technological changes over fifty years.

The districts are examined in order of their experience with imported technology. The Upper Great Lakes metallic-mining district, a true mining district, was a wholesale borrower of major, capital-intensive, technological innovations. The southeastern district of marginal forest farms, with its great mixture of metallic and non-metallic minerals, was equally a borrower and an innovator, but usually at a fairly unsophisticated and low-cost level of technology that was in keeping with its small-scale, mainly surface, operations. The western peninsula pioneer oil and salt district, with its prosperous farms and developing industrial centres, was in its day the most experimental and inventive of the three. And, while the Ontario petroleum fields ultimately proved to be low-yielding, the drilling and special refining techniques perfected there contributed significantly to new petroleum fields coming into production elsewhere in the world.

2

THE INNOVATIONS

During the nineteenth century, production technology underwent tremendous and varied changes on an international scale in response to the new industrial age. The second half of the century was characterized by rapid transformations from hand to mechanical technology in key areas of production; by an increase in the development of new basic technology; and by the spread of a relatively small number of similar production techniques to a large number of industries.[1] All these changes were evident in the field of mining, although no comprehensive history of them has ever been produced.

As late as 1850, serious mine operations everywhere were conducted largely by a hierarchy of skilled workers using traditional hand tools and empirical methods that had been perfected in Europe and practised relatively unchanged since antiquity, with some degree of mechanization becoming available from the sixteenth century onward.[2] Particularly significant in this context was the development in Great Britain in the early eighteenth century of steam pumping engines for dewatering mines. These enabled mine operators to reach the deeper-seated ores, which in turn induced experimentation with other innovations that facilitated deep-level mining and increased levels of production. In general, however, only rich veins and lodes could be profitably exploited using the traditional European mining methods.

The steady progress in economical application of machine power to mining operations generally was punctuated by technological breakthroughs after 1850. The United States came increasingly to dominate in the fields of mining equipment and engineering. By the end of the century, the requisite mechanically based technological innovations, plus many chemically based ones, had been worked out and applied to some degree in mining districts throughout the world. With increased mechanization, growth in the size and efficiency of machinery used, and greater quantity and velocity of throughput obtained in milling, it became possible to handle material at substantially lower costs per worker, per unit.[3] Moreover, in processing, including massive chemical treatment of ores, there occurred a substantial increase in the capacity to handle lower-grade and "rebellious" ores, and the same held true for the processing of inferior brands of petroleum.

In Ontario the awareness of, and initial receptivity to, the enormous technological changes in mining worldwide must be assessed in a dynamic framework of innovation and diffusion. The willingness of operators of Ontario mines even to test new ideas and machinery depended initially upon the availability of externally generated technology suited to their operations. The technological changes most applicable to mining in a newly developing territory like Ontario were new sources and means of transmission of power to operate mining activities, including pumping petroleum and brine wells; innovations in prospecting, drilling, and blasting operations as applicable to mining; and new processes of concentrating and separating mixed ores, and of refining petroleum and evaporating brine. An examination of technological changes in these areas will help determine which new techniques and machinery came into practical use, at what time, and in which forms they were introduced into Ontario mining operations. Because mining petroleum and salt required different techniques from those used in other minerals, the two types of mining are discussed separately.

PROSPECTING AND EXPLORATION

Metallic ores originate as massive inorganic orebodies that have been broken up and scattered or are deposited in fractures or fissures in the earth's crust in metamorphosed rocks. Ore structures are infinitely variable, ranging from "veins" or "lodes" to huge rock masses or blocks of more-or-less uniformly disseminated mineral. "Paying ore," as the name implies, refers to occurrences that contain high enough concentrations of mineral to be worked profitably under existing conditions. Conversely, "lean ores" are the poor ones, those which contain a lower percentage of metal than generally can be worked economically. Possible workable deposits had to be proved by determining the character, richness, and extent of deposits. These observations apply equally to such non-metallic ores as mica and apatite.

Explorations for mineral deposits prior to 1870 relied mainly upon geological skills and oral tradition rather than technological innovations. Workable deposits of minerals lying near the surface were found by testing surface showings or outcrops. Determining the physical properties of various minerals (hardness, specific gravity, and chemical reactions to blowpipe tests) aided in identifying them where this was not readily apparent. Assaying was undertaken to determine the grade or proportion of various metals in ores and to indicate whether a given metallic ore would repay treatment once it was mined. Only by clearing the surface and constructing test pits or by sampling ores in the shafts and tunnels of working mines, however, could operators actually ascertain the extent and shape of mineral deposits and decide whether further development was warranted.

That exploration and prospecting techniques of the day were highly labour-intensive and not especially practical or efficient is made clear by Morris Zaslow in the following observation concerning prospecting for economical mineral deposits that is still largely true of hardrock mining today:

> The task of finding a workable deposit, even in a known favourable area, entails hunting for hundreds of "showings" spread over a wide area; stripping away the overburden, digging and trenching to get an idea of their extent and trends; searching for an "occurrence" that might warrant further efforts to assess its size, content, grade, and shape by additional work . . . and sinking shafts and tunnelling that could raise the "occurrence" to the status of a "prospect." Hundreds of occurrences subjected to such efforts might yield only a single prospect; and even then, dozens of prospects are abandoned as too small, or too low grade to justify the capital outlays needed to develop them.[4]

Concern with reducing the time and risks involved in discovering workable deposits led the mining industry to develop innovations that would aid in detecting and delimiting subsurface mineral deposits, especially in the absence of any surface indications. Not until the twentieth century, with the founding of the science of geophysics, would serious headway be made.

The Diamond Drill

The major innovation in prospecting during this period was the diamond-pointed exploration drill, a hollow, fluid-circulating rotary drill that allowed operators to test the character and value of mines and quarries by running out core samples of strata from great depths. The core could then be analysed to indicate the size of the orebody and the amount of recoverable metal in each ton of ore. This was an innovation whose time was coming.[5] It was adapted from the first practical diamond-tipped drill developed by a French engineer, Rodolphe Leschot, in the early 1860s for driving the Mt. Cenis railway tunnel between France and Italy. The principle of employing industrial diamonds embedded in boring tools as ring cutters was simultaneously being experimented with in England.[6] Numerous improvements in 1869 by Holt and Severance, the American

Plate 1. Diamond Drill at Silver Islet, 1870s. *The major innovation in prospecting during this early period was the diamond-pointed exploration drill. In this sterograph by James Esson, a diamond drill is in use at the famed Silver Islet Mine on Lake Superior. The gentleman in the foreground on the right is William Bell Frue, inventor of the Frue Vanner and manager of the operation.*

diamond drill manufacturers, led to several specialized versions of the drill being developed, the most significant of which was a portable steampowered core drill for boring prospect holes.[7]

This diamond borer was a substantial improvement over Leschot's original machine and exemplified how a large number of small improvements were needed to bring a major innovation into practical use.[8] The improvements made the diamond drill highly suitable for mining, and word of it quickly spread throughout popular and professional publications.[9] Holt and Severances's improved diamond prospecting drill was tried successfully at the St. Joseph Lead Mine, Missouri, in 1869, and in Colorado silver mining in 1870.[10] It was brought to Ontario's Silver Islet Mine in 1873 by the mine's American owners, and in that same year it was employed elsewhere in Canada in geological survey work in New Brunswick, Nova Scotia, and in the North-West Territories.[11] In Ontario, a diamond-tipped exploration drill also was successfully employed in 1876 to explore the massive salt deposit at Goderich and in the Coe Hill iron ranges of the southeastern district later in the 1880s.

Geological Survey accounts of the contemporary use of a diamond drill of the same New York manufacturer offer a rare insight into the difficulties with its use that might explain why it was not instantly popular with the mining industry.[12] Its use for prospecting in the New Brunswick coalfields during 1873 and 1874, for instance, was plagued by every imaginable difficulty. Owing to the negligence of the supplier, the drill arrived on site with parts missing. After that difficulty was resolved, operators found it very complicated and sensitive to operate. Care had to be taken in starting the machine and also with the core bit, for if the diamonds fell out and became lost, a new supply could be procured only from the New York manufacturer. The machinery often jammed or broke down altogether because trained operators and mechanics to handle and maintain it were difficult to find and because water to cool the drill bit was not always available. Delays were also experienced because of the lack of materials to repair damages on the spot and the time lost in transporting the parts to the nearest town. Operators found that special hoisting equipment was required as greater depths were reached. The drill also proved less portable than had been hoped, and further delays resulted while a road was built to transport the drill to the spot selected.

Of even greater significance than these technical and locational difficulties was the great deal of experimentation and testing needed to reach a sound application basis, as will be seen in greater detail in Chapters Four and Five. The diamond drill could only be as effective as the skill and knowledge of the operators who were using it. More specifically, it was useful only in known mineral areas or areas containing anomalies, and this, along with its $6,000 price tag and its high operating cost, made its use uneconomical in most cases.

EXTRACTION

Mining consists of extracting mineral deposits included in solid rock formations by means of either open pits or trenches or underground openings.[13] If an orebody is large and near the surface it may be moved by open-pit methods – digging the ore from a large opening at the surface. Underground mining is always more complex, technologically demanding, and labour-intensive, and therefore it is the most costly of all. It entails sinking shafts or driving tunnels and lateral workings in a series of levels to gain access to mineral deposits; supporting the opening in unstable ground by "timbering"; then extracting (stoping) the ore. Stoping work was traditionally performed by hammer and drill men, who excavated the productive ore and dead rock at the working face by drilling and blasting, and muckers, who collected the ore and loaded it into carts. The workmen performing the auxiliary underground tasks in connection with sinking and driving – hauling and hoisting; perhaps ventilation; pumping; timbering; and routine mine maintenance – were referred to as "day men" or "labourers."

Power Transmission

Because mining was labour-intensive and artisan in nature and involved the application of human or animal energy, the trend throughout the nineteenth century was to multiply that energy through mechanical means. This consisted of applying new power sources and more efficient systems of power transmission to mining operations. Although this was the age of steam, the use of steam as a prime mover in underground mining operations was limited since the steam could neither be generated nor exhausted underground without fouling the atmosphere, nor could it be transmitted any distance without considerable loss of efficiency. It could be employed at the surface, however, to replace the traditional horsepower for hauling and hoisting purposes.[14] Steam was also used to pump out mines in the nineteenth century, and it was applied to power drills by the 1850s, but these were successful only in open pits and quarries.

The turning point came in the late 1860s and early 1870s when compressed air was introduced as the first efficient, practical means for replacing human and animal labour in underground operations.[15] Power transmitted by compressed air meant, in the words of an historian of American engineering, Robert Vogel, "the capability of transmitting centrally produced power to slave machinery and tools scattered indiscriminately over a ... site with utter flexibility of distance, location, and movement."[16] Compressed air was employed in England by 1863 for powering machinery for draining, hewing, and hauling coal. By the late 1860s it was successfully applied to machine-drilling of blast holes.[17] Then the late 1880s brought experiments applying electricity to mining operations in the United States, Great Britain, and France.[18] The most critical need for applying electricity to underground operations, however, was in coal mining, to overcome the fire hazard posed by coal gas.

Drilling

Following the discovery of a workable deposit, the task of mining began, a task in which rock drills and explosives were essential. A major focus of technological change in mining machinery was in the area of drills, which had two main purposes. The first was to secure samples of ore beyond the exposed rock face (prospecting). The second was to make an opening in a mass of rock in which to place explosives (blasting). A third related but different drilling operation was boring artesian wells. Sinking of pits and shafts and driving tunnels usually required the removal of substantial quantities of useless rock,

Plate 2. Miners at Silver Islet, 1870s. *Underground mining was technological demanding and labour-intensive in the era before the full-scale use of mass-mining techniques. Depicted in this photograph are hammer and drill men. They excavated the productive ore and dead rock at the working face. The boy with them likely worked as a sorter in the rock house.*

referred to as "waste" or gangue. From earliest times removing dead rock and excavating ore was done by the ancient technique of firequenching and wedging or by pick, hammer, and chisel.[19] By the eighteenth century, most mining operations required hand-drilling shot holes, then blasting the rock wall with black powder explosives. The need to increase the speed, efficiency, and, as a consequence, safety with which rock was removed led to much experimentation.

Many improvements in the design of a practical rock drill resulted from the search for a mechanically powered, light, compact, simple, durable drill capable of boring true and deep shot holes at any angle. The first experiments were conducted for the field of tunnel engineering, not mining, and involved the use of steam.[20] Most of the new rock-drilling machinery bore by percussion (drop drills), but there were also ones which bore by constant pressure and rotation. Historians claim that the Fowle Rock Drill, developed in the United States in 1849 and patented and used in quarrying by 1851, was the first workable machine drill in that country.[21] Despite its practicality for rock-drilling generally, however, the Fowle drill was impractical for tunnel work because, like other machine drills of the day, it was steamdriven. Only when the operating principle was combined with such other innovations as compressed air did a truly practical rock drill for tunnel work and mining arrive on the scene.[22] The first workable application of compressed air to rock-drilling was the Sommeiller Drill, developed in 1861 to drive the Mt. Cenis Tunnel.[23] Air was compressed at the tunnel portals and piped to the drills at the workface, something which would have been impossible to accomplish with steam. Though the drill was complex and highly unreliable, the fact that it was driven by air, rather than steam, was portentous.

Once again the railway tunnel projects led the way to technological innovations. A revolutionary new compressed-air machine drill successfully combining piston-action and automatic drill rotation was developed in 1865 by Charles Burleigh, a Massachusetts machinist, to speed completion of the Hoosac Tunnel.[24] After a few years of experimentation and further improvements by Burleigh, his drill quickly became the prototype of all piston-type drills. At the Hoosac, several drills were mounted on a strong, wheeled framework, and the whole was operated as a "battery." The English simultaneously conducted experiments with similar piston-type, self-advancing, compressed-air drills.[25] Burleighs were introduced at the Silver Islet Mine by the American mineowners very shortly thereafter, in 1872. The use of compressed-air drills spread quickly throughout the entire northern district of Ontario over the next ten years. The drills were only introduced in the southeastern district during the 1880s, and even then in a single instance, when a serious attempt was made to undertake full-scale development of an apatite property near Kingston. This observation will be explained in Chapter Five. The drills were expensive, costing approximately $250 each in Ontario. The pipe and small compressor cost an additional $3,500.

The particular compressed-air drill adopted in North American mining camps was not the Burleigh, but rather improved versions of it, most notably the Rand and the Ingersoll.[26] The straight transfer of the Burleigh drill from construction projects to underground mining was a failure, as historian Larry Lankton has convincingly demonstrated.[27] The peculiar and difficult circumstances of mining were hard on power drills. Underground openings were small and irregular, the atmosphere was damp, and the drills had to be moved about constantly. The regular Burleigh drill was complex, which tended to make it too bulky for use in small spaces and also created repair problems. Moreover, as mentioned previously, the operating requirements of the Burleigh threatened the independence of the miners who used it so they resisted its adoption. Improvements in the 1870s and early 1880s produced drills that were smaller, lighter, and more durable, yet still powerful, while improved columns, stands, and frames were developed that stabilized the mechanical elements and enabled the drill to be worked in any conceivable position.[28] Of all these features, however, the question of machine repair problems was an all-important one in remote areas like the Upper Great Lakes district of Ontario.

The other major innovations in tunnel-drilling besides the piston-drill operated by compressed air were the concurrently developed diamond-pointed rotary drill, already mentioned in connection with exploration work, and the elecric drill. Adaptations to the diamond drill were speedily developed for mineral extraction; as early as 1869 Holt and Severance were marketing special diamond drills for quarry and tunnel duty.[29] The English version of the machine was employed at the Spring Hill Collieries, Nova Scotia, in 1873 and was patented in Canada in 1875.[30] No record exists of its use in Ontario mining (other than for purposes of exploration) in the period before 1890, however. The first practical electric drill to appear on the market was the Edison General Electric Percussion Drill, patented in 1891, just outside the period of this study.

Incidentally, drill development from hand to electrically powered followed no single line. Rock drills were made more effective, regardless of how they were to be powered. The trend, in fact, was to develop drills capable of being worked by a choice of either hand or power.[31] A mine operator who employed hand drills in the 1870s and 1880s may in fact have been using relatively advanced drilling equipment.[32]

Blasting

The ultimate goal in improved blasting techniques was to reduce the number of blast holes required over a given working face and to bring down more rock per blow by drilling deeper blast holes. This was a delicate business at best. The most practical and logical direction for improvements, besides using more powerful and efficient machine drills for making the blast holes, lay with devising reliable fuses and detonating apparatus and developing powerful blasting materials that were safe to use. Black powder (gunpowder) and the Bickford slow-match fuse, a reliable and flexible fuse developed by a Cornishman in 1831, were employed exclusively for blasting until European experiments with nitroglycerine in the 1840s led to the development of more high-powered explosives.[33] Nitroglycerine was a more powerful explosive than black powder, but it was also a highly volatile liquid that had an unfortunate tendency to decompose with heat and age, making it dangerous and, hence, undesirable to use.[34]

Nitroglycerine was first introduced in the United States in 1866 on the Hoosac Tunnel as being far more effective than black powder for blasting underground because of its reputation for great force and its virtual smokelessness.[35] To help obviate the dangers attendant on its use, the explosive was substantially improved by an English chemist, George W. Mowbray, who had been commissioned to develop techniques for its bulk manufacture and safe employment at the Hoosac.[36] Mowbray's main contribution lay in systematizing the use of nitroglycerine. He refined the manufacturing process and froze the liquid to improve the explosive's stability and developed a new fuse and a static-electric blasting machine that produced simultaneous detonation and so increased the effectiveness of each blast.[37] At the same time the famous Swede, Alfred Nobel, introduced an even more stable, solid form of nitroglycerine, called "dynamite" or giant powder. A number of commercial variations came on the market – Dualin, Mica pwoder, Rendrock, Vulcan powder – all of them simple types of solid nitroglycerine compounds.[38] Nobel also continued experimenting with an improved explosive at his plant in Glasgow. By the mid-1870s, he had successfully converted the liquid nitroglycerine into a safer, gelatine form.[39]

The next major breakthrough came around 1880, with the manufacture of the Abel water cartridge, or waterproof envelope, which Nobel combined with his gelatine explosives (gelignite and gelatine dynamite), a new fuse, and an improved system of detonation to produce the Settle Gelatine Water-Cartridge for safety blasting in "firey" conditions.[40] It is not known when nitroglycerine was first introduced in Ontario mining operations, but Mowbray himself established an explosives plant at Kingston about 1876.[41] Moreover, a patent for the manufacture of nitroglycerine had been granted to C.W. Volney, a Brockville manufacturer, in 1873.[42] From the mid-1870s onward nitroglycerine in cartridges was used in conjunction with drilling or rejuvenating Ontario petroleum wells; and the Dualin version was widely used in the southeastern district in the 1880s for all types of mining. In contrast to most of the new mining technology, there were no capital barriers to the spread of this technique, so its diffusion could be expected to be rapid and complete.

CONCENTRATION AND SEPARATION

Extracting the ore and waste rock was the last operation in mining to be mechanized, while concentrating and separating had been the first. The ore brought to the surface was in most cases a mixture of particles of the desired minerals with other minerals, plus varying amounts of waste rock. The desired mineral matter had to be separated from the dead rock, and, if possible, the desired minerals had to be separated from one another. Metallic ores were treated by two types of processes: the first process was a mechanical one; the second was chemical. Mechanical processing involved breaking ore by crushing or stamping to free the desired metallic minerals for concentration and sizing in a mill, the capacity of which in the period before 1890 did not exceed about five hundred tons per twenty-four hours. Milling separated the various minerals by removing as much valuable matter as possible; often by a complex sequence of mechanical operations. The work required continuous shifts of unskilled and semi-skilled labourers working under fairly close supervision. Mills were almost always located at or near the mine as it usually proved cheaper to concentrate the ore at the source in order to eliminate as much waste material as possible before shipping. Finally, smelting and refining processes were necessary to convert the mineral concentrates and iron ore, which did not require concentration, into pure metals. As these last steps typically involved chemical and thermal processes and required considerable amounts of cheap fuel, they required highly skilled workers. They were also more likely to take place close to sources of fuel and skilled labour and to markets than at the mine site. An additional advantage for the operators of a centrally located custom smelter was their ability to buy and combine for smelting a variety of ores to their profit.

Milling

Crushing and Grinding

From the standpoint of a metallurgist, ore may be defined as a mineral aggregate containing metal, or metals, in sufficient quantity to make their extraction commercially valuable. A particular species of ore is named for its dominant metal. Ores of copper, iron, lead, and silver were hand-sorted and graded as they came from the mine. Those which contained a high percentage of metal, the high-grade ores, were sent directly to the smelter. Low-grade ores always had to be dressed to remove the worthless gangue in which the metals were trapped; these ores were known as "milling ore" or "stamp work," and the valuable heavy part obtained from them was called "a concentrate." Free-milling gold and silver ores were slightly different since they were processed according to their character in custom mills in which the ore was crushed and the metal extracted directly from the pulp by means of amalgamation. Overall, the ore concentration devices employed at any mine depended largely on the nature of the mineral, the characteristics of its occurrence in a given setting, and the scale of production.

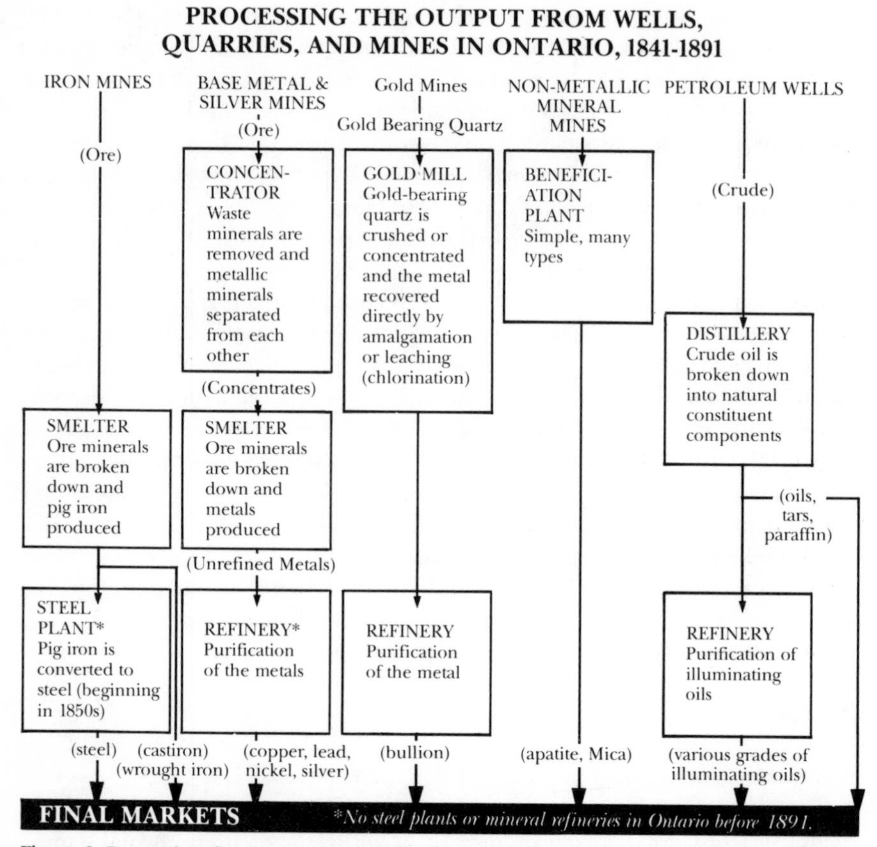

Figure 3. Processing the Output from Wells, Pits, and Mines in Ontario, 1841-1891. *Smelters and refineries (except for petroleum) were rare, and steel plants non-existent, in Ontario prior to 1891.*

The primary objective, which determined the nature and elaborateness of the concentration process, was to obtain the highest economic recovery. To this end attempts were made to mechanize the equipment and to increase its capacity and economy. In addition, there was a trend away from using gravity feed and discharge, and batchwork, toward introducing graded crushing, graded separation, and specialized machinery and plants that treated a continuous flow of materials. The desired effect of this trend was to reduce also the number, skills, and supervision required of the mill hands.

The first step in milling was comminution, breaking down the ore into smaller sizes by crushing and grinding it until each useful particle was separated from the waste material. The methods of crushing and grinding had been partially mechanized as early as the sixteenth century, and these have been carefully detailed in Agricola's well-known work, *De re Metallica.*[43] Traditionally the crushing was done by hand or animal-powered rotary millstones, or in larger plants by waterdriven pestle stamps, or, by the nineteenth century, by rolls. Developments in crushing after 1850 involved applying steam to power larger capacity, more strongly built equipment, American improvements to the Cornish stamps and Cornish rolls, and the American invention of jaw crushers.

Cornish stamps, which were Saxon in origin, were introduced to mining in Cornwall in the seventeenth century. These were gravity stamps, lifted by cams, which woud drop under their own weight to strike the ore. These stamps were completely remodelled at the hands of the millmen in the 1850s in the mining camps of California. With the California stamps, parts were interchangeable, the heads rotated to wear uniformly, and overall the apparatus was more durable and of greater capacity than the traditional stamps.[44] The 1850s also witnessed the introduction of steam stamps, whereby the descent of the stamp was caused partially or entirely by some mechanical force, which were developed simultaneously in England and the United States.[45] Steam stamps made it possible to crush large tonnages of ore in a limited space. The most important early American developments occurred in the Lake Superior copper district. First, there was the Ball stamp, in which were combined the principles of the Cornish gravity stamp and the direct action steam hammer. It was adopted throughout the copper district immediately upon its invention in 1856 by William Ball, to be superseded about 1879 by the Leavitt steam stamp, a substantially more efficient and economical version of the Ball stamp.[46] The Leavitt was developed in the late 1880s at the famed Calumet and Hecla Mines, by the consulting engineer E.D. Leavitt, Jr. The Krause steam stamp followed in 1888.[47]

Stamp mills were so basic to concentration and the small waterpowered versions were so easily and cheaply erected and repaired that their use in mining after 1850 was practically universal. Even the most marginal hardrock operations in the Ontario districts erected a small (two- to five-stamp), gravity stamp battery as soon as mining began. The copper-, silver-, and gold-mining operations in the Upper Great Lakes district employed much larger, steamdriven stamps; the Silver Islet Mine alone employed eighty Ball stamps at the peak of its production, and the Badger Silver Mine employed Krause stamps in 1890. The cost of a battery of steam-stamps amounted to roughly $1,000 per stamp.

Stamp mills had a second different but related use – crushing and amalgamating free-milling, oxidized gold and silver ores. Ore was stamped fine, and the pulp run over amalgamation plates located inside or near the stamp mortar. Gold particles adhered to the amalgamation plates and also fell to the bottom of the mortar, where they were caught by a mercury-trap. Afterwards, the gold was "cleaned-up" and retorted. A proliferation of these quartz crushers appeared on the North American market during the 1850s and 1860s.[48] Most of these innovations involved gravity stamps, but also there were steamdriven ones, such as the experimental Forbes Quartz Crusher. As will be seen in a later chapter, the Forbes Crusher was an American patented invention which was tested and rejected as uneconomical at the Dean and Williams gold mine in the Marmora area in 1872.

Seldom could stamps be employed without the use also of rolls and crushers, and in many circumstances, such as in milling the copper ores found in Montana, rolls and crushers replaced stamps altogether. Cornish rolls, an innovation of the early nineteenth century, comprised twin iron cylinders sheathed with chilled cast-iron. The rollers, measuring only a few feet in diameter, revolved slowly, pressed together by powerful weighted levers. The improved Cornish rolls adopted widely in North America after 1860 were the Krom rolls; these featured larger reduction ratios and were more strongly built, more compact, and ran at higher speeds than their prototypes.[49] Traditional Cornish rolls and Krom rolls were both standard equipment in the mills of the Upper Great Lakes district by the 1870s.

Apart from these improvements to the Cornish crushing machinery, there were a number of entirely new crushers which also acted on the principle of gradual pressure. A nutcracker-like jaw crusher, named the Blake Rock Breaker, was developed in Connecticut in 1854 for road construction and was in use almost universally in major mining districts throughout North America by the 1860s. The Blake crusher was employed for rough crushing preliminary to stamping. Unlike the Burleigh rock drill, the Blake crusher was so efficient from the beginning that it was not superseded until well into the twentieth century.[50] The success of the jaw crusher did not preclude further experimentation, however, as a number of gyratory (or rotary) crushers eventually came on the market as well. These were of larger capacity than the jaw crushers. The most popular one was the Gates, developed in the Michigan copper region in the 1880s and marketed by the Fraser and Chalmers Company.[51] Its use in Ontario is documented for the Badger Silver Mine in the area of Thunder Bay.

The next stage, grinding, was accomplished by rotating, fine-grinding mills which operated on the principle of attrition, percussion, or a combination of the two. These grinders were often all-purpose, useful for cereals, paints, and similar products. The most traditional of the roller mills was the arrastra, which employed the ancient, simple process of dragging heavy stones across a pit filled with crushed ore to be ground; and the Chilean mill, similar to the arrastra but using a millstone set vertically to roll around an annular trough.[52] These two traditional methods of grinding underwent substantial improvements in the 1850s and 1860s to make them more useful for grinding ores.[53] Because they could be powered by water or animals, these mills were especially useful for treating ores in the early stages of a mine's development or in remote areas. Special improved versions of grinding pans were extremely popular for fine-grinding of gold or difficult silver ores, the object being, however, the chemical operation of amalgamation of concentrates rather than the mechanical action of crushing.[54] Some grinders, called "pulverizers," involved the percussive effect of weights against a revolving track, the weights being impelled either by centrifugal force or gravity.[55] The weights themselves took the shape of balls or rollers. Fine grinders which combined the principles of percussion and attrition were known as "disintegrators."[56] In Ontario, pulverizers of unknown origins were used in the 1880s in a few silver-mining operations in the Upper Great Lakes district, and a Carr Disintegrator was used in 1871 in the Brockville Superphosphate Works in the southeastern district. The ultimate improvements in fine-grinding mills would be after 1890, with the introduction of tumbling mills: the ball mill and, later, the rod mill, both of which still can be found in modern concentration mills today.

Specific Gravity Separation

Grinding reduced the ore to about the particle size of salt or sand, thus freeing the minerals from the host rock. Once liberated, the minerals had to be separated from the waste and, it was hoped, from each other. The important separations were those according to size and chemical composition and of solid from fluid. Nineteenth-century improvements to mechanical separation techniques involved some experiments with magnets (for magnetic ores), but the bulk of the experimentation was with specific gravity separation.[57] Indeed, until the development in the early twentieth century of froth flotation, differences in the specific gravity of minerals formed the basis for all concentration processes.

Put very simply, specific gravity differences between minerals cause differences in their behaviour when settling in water, air, or other fluids. The separation of valuable minerals – usually the heavier part of the ore – from the waste was rarely accomplished in a single step. The most popular method involved sedimentation in fluids. Until the 1850s, washing to separate the mineral particles from the worthless gangue was done by hand sieves, troughs, and jigs. Separating the mineral particles of varying densities was done in more elaborate set-ups: in pans (or buddles), frames, or tables, all of which followed traditional Cornish methods.[58] Because of the trend towards increasing the throughput in the last half of the nineteenth century, millmen sought to introduce continuous process machinery and plants that would increase the overall separation efficiency, capacity, and velocity of their separators and concentrators. They sought also to improve the methods for the economical recovery of fine particles of minerals from the finest-ground material, called "slimes," and from the mill tailings.

Coarse-sand concentration, the first to be mechanized, was accomplished by machines such as the Collom Washer and the Edward Washer, both of which were developed in the 1850s for the Lake Superior copper district, and jigs.[59] The Collom washer was used in several mining operations in the Upper Great Lakes district, and the equally popular Hartz Jig received a trial at a leading Marmora-area gold mine. These early mechanical washers and jigs usually only added to the length and cost of the washing process without producing any appreciable savings for the mill owners.

The most spectacular of all the technological breakthroughs in ore separation and concentration to occur before 1890, however, was found in the area of fine-sand and slime gravity separation. While the designs of the individual machines varied considerably, each type took advantage of the principle of particle friction in relation to movement on a surface, the movement being caused by applying a horizontal current of water, often supplemented with external impulses such as bumps or shakes. For instance, the automatic revolving slime table was developed to improve upon the traditional Cornish buddle. In this regard, a conical slime table, named the Evans, was a popular local invention of the mid-1870s to resolve the particular problems with milling the conglomerate rock of the Michigan copper region.[60] Then there were the shaking and bumping tables, most of which were continuous and self-acting.[61] Last were the vanners, which differed from the revolving, bumping, or shaking tables, with their stationary surfaces, in that the concentrates travelled on the upper surface of an endless, inclined, belt of canvas stretched around a strake frame. The vanner was probably based on the Bruton ore-dressing machinery, or "Bruton's Cloth," introduced in 1842 in Scotland for the recovery of lead concentrates from slimes.[62]

The vanner is of particular significance in the history of Canadian mining because it was perfected in Ontario. The ultimate improvement was the Frue Vanner, which was developed in Ontario at the Silver Islet Mine during 1872-74 by its captain, a Michigan mining veteran, William Bell Frue, and his master mechanic from Fredonia, New York, William Foster.[63] With the Frue Vanner a secondary agitation, a side-shaking motion, was added to the forward crawl of the belt – oldtimers have been known to call them the "hula dancers." This added motion, plus small substitutions and adjustments of Frue's original design permitted a 90 per cent recovery from slimes previously thought worthless. Following his invention and its successful use in the Upper Great Lakes district, Frue was hired and his patents purchased in the 1870s by Fraser and Chalmers Company of Chicago. At a cost of roughly $750 the Frue Vanner was not a cheap item. Yet, the economic advantages from using the Vanner, plus its versatility, were so apparent that the rate of its diffusion throughout the mining world was rapid.

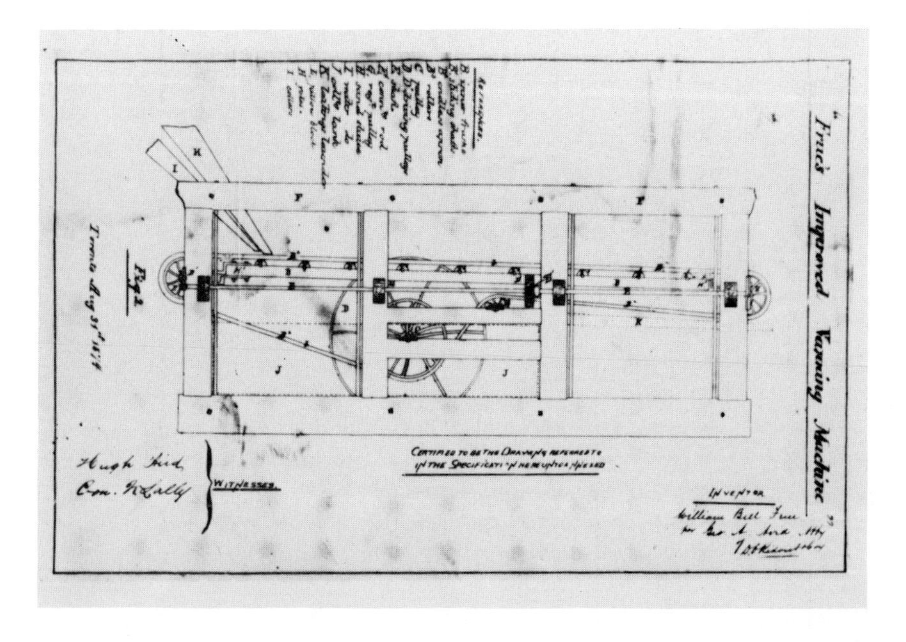

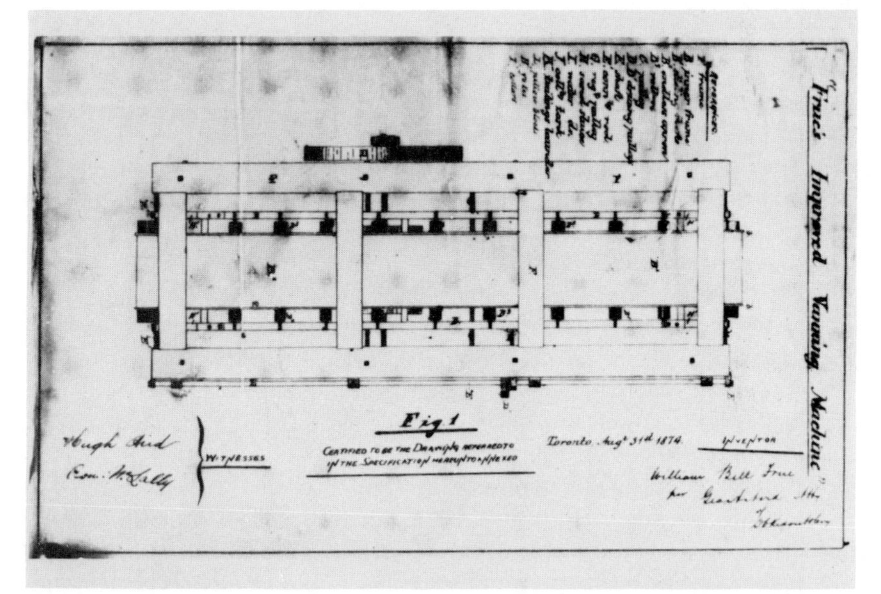

Plate 3. The Frue Vanner, Patent Drawing, 1874. *This patented invention by the Silver Islet company manager, William Bell Frue, eventually enabled the company of operators to work the lean ores profitably. It was the final improvement to a long line of developments in mechanical ore separation devices. Its main contribution lay in its high recovery of fine particles of metal from the finest-ground material, called "slimes," and from mill tailings. In addition to its impact in the Lake Superior section, the Vanner proved also to be a considerable long-term contribution to the large world of metallic-mining.*

Smelting and Refining

The ultimate process by which most useful minerals were broken down and other elements recovered was a chemical one, smelting, that required a highly skilled workforce. Prior to smelting, however, most forms of metallic ores required preliminary treatment by roasting or calcining them at temperatures below their melting points. Roasting, which was very labourious, hot, and exhausting work, changed the sulphide and carbonate ores to oxides or smelting ores. When roasting sulphide ores in lump form, the time-honoured technique of heap-roasting at the minesite was the most frequently used. Production was extremely slow with heap-roasting, however, taking one to three months, not to mention a vast quantity of cordwood and a great deal of hand labour, to roast a single batch of several hundred tons. When ignited, dense sulphurous fumes were given off. All these drawbacks notwithstanding, heap-roasting was the most economical method to use in a new, untried mining district where the mine owners would be reluctant to invest prematurely in a physical plant.[64] Lump ores containing carbonates required roasting in small, open-shaft furnaces, or kilns, because carbonates, unlike sulphides, will not burn and therefore have to be blended with the fuel. In Ontario, heap-roasting occurred in the Sudbury Basin in the late 1880s.

With pulverized ores and concentrates, the methods of roasting were somewhat different than with lump ores. First there was the sixteenth-century Welsh process, whereby skilled furnacemen subjected concentrates to successive oxidation in a series of reverberatory furnaces.[65] In this case, where the hearth is kept separate from the fire, the products of combustion and the flame are drawn over the hearth to the chimney, and an arched roof over the hearth reverberates the heat down upon the charge. Because moderate temperatures were required and little heat was lost through radiation, the reverberatory furnace was economical on fuel. The only serious objection to the Welsh process was the cost of the large force of skilled labour it required. Nevertheless, in the early days of mining in North America, when the entire workforce had to be imported anyway, reverberatory furnaces were a popular form of roasting. A Welsh roasting and smelting works was installed in the Upper Great Lakes district at Bruce Mines in 1849.

American improvements to reverberatory roasters after 1850 resulted in the mechanically operated, compact, circular furnace having a continuous feed and discharge. The most significant achievement in this area before 1890 was the revolving cylinder furnace with the axis horizontal or inclined toward the discharge end.[66] These furnaces where charged and discharged through manholes, and the inclined furnaces permitted the pulverized ore to travel gradually from the feed end to the lower end mechanically as the furnace revolved. Therefore, it required fewer skilled workers to tend to it. An inclined, revolving cylinder furnace that incorporated a great deal of the new technology was experimented with and patented in 1872 in the goldfields of southeastern Ontario.

In the smelting process proper, metal oxides were subjected to melting temperatures and gave up their oxygen to become metals. Carbon fuel (charcoal, coal, or coke) was used to supply the heat and also to act as the reductant in the chemical reaction. Two quite different processes using this chemical reaction had been used for centuries to produce copper, pig iron, lead, and tin. The first of these was the sixteenth-century Welsh process of reverberatory smelting, which was similar to that of Welsh roasting. It consisted of treating copper ore in a series of roastings and fusions.[67] The advantage of using the reverberatory furnace for smelting copper lay with its efficiency with fine concentrates, with ores from lodes that changed suddenly or greatly, and with ores of very high grade. Reverberatory furnaces also were economical for smelting in regions like the Upper Great Lakes, where it was necessary to use cheap fuels. While there occurred numerous changes to improve the capacity and economy of reverberatory smelters, the basic practice remained unchanged.

The other method of smelting was in blast-furnaces, which had a larger capacity and were more economical to operate than the Welsh reverberatories.[68] It followed the traditional German technique of smelting copper and silver-lead ores in a blast-furnace charged with an intimate mixture of ore, reductant/fuel, and flux. Here, the charge was laid on the hearth, ignited, and, by means of a forced draft or "blast," heated to high temperatures, whereupon it was reduced to molten metal which collected at the bottom of the furnace. In contrast with the reverberatory furnaces, blast-furnaces underwent a major transformation at the hands of Europeans and, more spectacularly, Americans.

The major American advance over the old German process for smelting was the water-jacketed blast-furnace, which was developed in the 1870s and perfected over the next several decades. With this new furnace, cold water circulated between inner and outer shells, preventing the inner shell from being melted or attacked by the highly heated contents of the blast-furnace. This represented a significant advance because it permitted the intensification of energy while at the same time it eliminated the burning out and "freezing up" of half-fused masses of molten firebrick – a phenomenon which occurred frequently with the old blast-furnaces. This type of furnace could both roast and smelt, and it required less fuel per ton of ore than a reverberatory. Despite the obvious advantages of the water-jacketed furnace for smelting, it was better adapted for the treatment of coarse ores than of fines and concentrates. And, the costs to build one were so high – it cost $120,000 for a typical water-jacketed furnace of 100-ton capacity – that only the largest, well-financed, operations could consider erecting one.[69] Its first use in Ontario was for processing the nickel-copper ores of the Sudbury Basin in 1889.

With lead-bearing ores, high-grade concentrates were easily made using fairly crude appliances. This was so because lead-bearing ores are easily milled and lead has a very low melting point. The simple process of smelting only for the lead content of high-grade ores was accomplished at first using small reverberatory furnaces, which by the 1850s were replaced by an American innovation, the so-called "American Ore-Hearth."[70] The American ore-hearth was little more than a highly simplified and modified version of the old German blast-furnace. It received at least one trial in Ontario, in the mid-1860s at the southeastern district's most promising lead mine, the Frontenac.

Improvements to blast-furnaces for smelting iron ore and for steel-making have been the subject of so much writing that they require only a brief mention here.[71] Iron ores were smelted in a blast-furnace filled with a mixture of wood charcoal or coke, iron ore, and limestone for flux. A blast of air was blown in at the bottom; the fuel burned to maintain a melting temperature in the furnace and also to reduce the iron in the ore to metallic form as pig iron. Pig iron could then be cast, wrought, or converted to steel. Iron-making using either the cold-blast technique or the newer hot-blast technique, developed in the 1820s in Scotland, was experimented with in a limited way at a few locations in the Marmora area between 1820 and 1860. By the 1860s, the Bessemer steel-making process and, a decade later, the open-hearth technique, had been adopted in the United States. By 1890, the Bessemer process had been improved to the point where it was firmly established on a large scale in that country. But there were no steel plants in Ontario before 1890.[72]

Hydrometallurgical Treatment

Where smelting was too expensive and amalgamation was unsuitable, copper, gold, lead, and silver could be obtained by taking the metal into solution, then precipitating it from the solution. The gold-bearing quartz and silver ores which contained arsenic and antimony were refractory ores and could be treated only by hydrometallurgy. This led to the improvements in the western United States in the 1860s of the German chlorination process for treating gold.[73] The basic process involved roasting the gold-bearing quartz and subjecting it to bleaching powder or chlorine gas to form a soluble chloride from which gold could be precipitated with any base metal. Chlorination mills sprung up throughout North America in the 1870s, and in the 1880s they were tried in the Marmora and Lake of the Woods goldfields. But, as Otis Young points out in his history of Western mining, chlorination proved to be better in theory than in practice.[74] Not only were the chlorination vats and barrels expensive to construct and maintain and the chlorine expensive, not reusable, and lethal to its handlers, but also any base metal present in the ore would react with the chlorine before gold would; therefore, the method simply did not work to recover gold in most cases. The true resolution would be found only with the cyanide process, discovered in the late 1880s in Scotland and subsequently improved upon in gold- and silver-mining districts of Australia and the United States.

Chlorination was not only impractical for the treatment of gold quartz, it was completely unsuitable for the treatment of silver ore. Thus, a host of solvents and precipitants were experimented with specifically for extracting silver, the specific solvents and precipitants varying according to the form of the silver ore being treated. In North America, these experiments took place in the western mining camps, and it would appear that the various wet processes from there did not spread to the silver mines of Ontario before 1890.

In the case of extracting copper from low-grade copper ores, especially siliceous, and therefore highly unstable, ones, leaching was considered to be more economical than smelting as early as the 1860s. A salt process for leaching copper developed in the early 1860s in Scotland, for example, involved crushing, roasting to expel the sulphur, salting, and re-roasting the copper ore.[75] Then the roasted, chlorinated material was leached in a salt solution to extract the copper as a soluble chloride. Following that, the copper was precipitated on scrap iron, melted, and cast into ingots. This process was experimented with briefly in the mid-1860s at Ontario's Bruce Mines.

SALT AND PETROLEUM

Salt and petroleum operations involved quite different tasks from those just described and brought into being an entirely new set of skills and occupations. These mineral resources were produced by sedimentation, usually at the bottoms of shallow seas. Petroleum, known at first as rock oil, is a liquid hydrocarbon formed from decayed organic matter in some fossil-containing rocks that are overlaid by subsequent strata. Petroleum and natural gas – which are not, strictly speaking, minerals at all but which are usually thought of as mineral fuels – originated mainly in the Palaeozoic era. Being mobile and low in density, these materials were able to migrate from their points of origin, often concentrating in the folds or anticlinal arches beneath impermeable strata.

The escapes of rock oil were of two types – "gum beds" where the lighter oils had evaporated leaving the heavier, tarry materials, and surface springs. These two forms of seepages were first tapped in Ontario and Pennsylvania by wide-bore water wells in the soil overlying the bedrock. Others were sunk to greater depths in the rock layers, whereupon spontaneously flowing subterranean springs were struck. The deeper-lying strata were tapped by boring with existing artesian well techniques, a practice begun early in the 1860s. Before 1890, however, depths were measured in hundreds not thousands of feet. Once reached, the crude petroleum was extracted by pumping. Because crude retained its liquid form and was highly flammable, it required special storage and handling. It was stored in temporary tanks at the wellsite, collected and perhaps stored again at a more central location, packaged in wooden barrels and shipped in wagons to a refinery for separation into its various components, or "fractions," by distillation. The illuminating oils – the principal commercial use for petroleum before 1900, when it became valuable as a major source of energy – received further treatment for quality in a refinery.

Since salt beds usually occur in conjunction with petroleum and natural gas, the discovery of one of these three substances was likely to occur in the course of exploring for the others. Salt could be mined from its beds as rock salt, but an economical method of extracting in deep-seated beds was to convert it into an artificial brine by dissolving the deposit with water, then pumping or forcing the brine to the surface, and reconstituting the salt through the process of evaporation. Commercial preparation of salt also included such operations as crushing, removing impurities, grading, and packaging. Extracting salt as brine was the common method in nineteenth-century North America.

Drilling and Processing

Ontario's petroleum and salt-mining industries inherited the full range of techniques developed over the centuries for boring and operating artesian wells. The mechanical techniques developed in France in the eighteenth century had their roots in the ancient Chinese system of deep well boring that employed a rope carrying tools.[76] That system was capable of boring deep wells – one even reached three thousand feet – but it was not practical for hard strata. Conversely, the system developed by the English and Germans was not practical for deep borings. It comprised a series of iron rods jointed or screwed together and carrying various tools, which penetrated the strata propelled either by a rotary or percussive movement.[77] A major obstacle was the removal of debris as the bore progressed. The Kind system, created in France in 1842, for example, employed an elaborate series of complex operations that proved impractical. Improvements in deep-boring apparatus constantly sought ways of removing debris without at the same time having to remove the shaft or change the tools.[78] These European deep-drilling methods were combined and improved upon in the early nineteenth century by the highly skilled saline drillers of the Great Kanawha Valley in West Virginia.

The method of sinking the first petroleum wells, beginning in 1859, was to dig and crib a four-foot shaft down to rock, then sink the well through the rock using a succession of tools and a percussion action. At first wells were "kicked down" by manpower using spring poles. But this method was quickly superseded by steamdriven drilling rigs operated by specialized, itinerant teams of drillers.[79] The tools and the method employed in percussion drilling were basically those that had been developed several decades earlier by the saline drillers of West Virginia, which had then been adopted in the early 1860s by the oil drillers there and in Pennsylvania and Ontario. The skills in drilling increased and the machinery as applied to petroleum was improved upon, though the basic techniques remained the same. The system consisted of a bit or cutting tool in the form of a chisel screwed to an auger stem. This in turn was connected to "jars" or slips, which permitted a delayed power motion on both the upward and downward strokes, and then to a sinker bar for weight. The free end of the sinker bar was a rope or screw socket from which a rope or pole ran to connect it with the source of power at the surface.[80] As they were drilled, the wells were lined with lengths of wooden tubes or wrought-iron casing, screwed together in sections. The special needs of petroleum operations were to drill wells with speed, accuracy, and economy to much greater depths and through disparate strata. Deep wells required correspondingly heavier tools and greater sources of power.

In Pennsylvania, where the deposits ran deep, the direction of innovations in petroleum drilling was to improve the cable drill. This consisted primarily of replacing the rope with steel cable – itself a major American innovation. This improved cable-drilling system on the percussion principle became known internationally as the "American System."[81] The advantage of the cable system lay in its ability to bore to deep levels since the tools could be removed quickly. These developments contrasted with the situation in Ontario where petroleum was found at shallow levels and iron and steel were scarce and costly materials. The innovations here took a different direction, and a rotary drilling technique of solid wooden rods was introduced to the Ontario oilfields at Petrolia about 1866 to replace cable-drilling. Lengths of hardwood were screwed to each other during the course of drilling, being lowered into or withdrawn from the hole by means of a tall tripod erected over the well. The use of poles, as already mentioned, had a long history in drilling artesian wells since the poles could hold drilling tools together better than a rope or cable while boring through difficult rock and bore straighter holes.[82]

As will be discussed more fully in a later chapter, the man who introduced this pole technique to the Ontario petroleum industry was a local businessman by the name of William H. MacGarvey.[83] In brief, the components of MacGarvey's system, the wooden rods, down hole tools, locomotive-type boiler, and portable steam engine driving a band wheel, with a crank and pitman to drive the walking beam and hoisting drum, were largely conventional. The innovative feature lay in the particular pattern of their combination and in small design details, such as the wing guide and reamer, which MacGarvey added to them. It was widely adopted by the southwestern Ontario petroleum drillers and by many in the new oilfields opening to development elsewhere in the world, in eastern Europe and Mexico, for example. Its improved version was known as the "Canadian System" of drilling. The cost of a complete Canadian drilling rig manufactured at Petrolia in 1890 was a modest $1,700.

Once the well penetrated the zone of petroleum and gas, the hole had been cleared, and the lift pump put down, the well went into production. Within the first few years of discovery, the oilfields of Pennsylvania and Ontario were being worked by means of an ingenious, inexpensive system introduced to pump dozens of wells with a single small-powered steam engine. The method, known as the "jerker system," involved subdividing and distributing the power generated in a central power plant across a site to various points of application.[84] The whole pumping system was constructed almost entirely of wood. It was arranged so as to be in balance, half the dead load of rods and mechanism in the field being lifted while the other half was descending. The power required was only that for overcoming inertia and friction, plus the weight of the oil lifted at each stroke. The system worked in this way: the engine was connected to main or master wheels imparting a reciprocating motion to them which caused them to swing through an arc to give jerker rods the necessary forward and backward motion to produce a stroke at the pumps at each well. The rods, which were lengths of wood with spliced joints, were attached to the free end of a wooden walking beam or metal triangle arm; at the other end of the connection hung the pump rod.

Plate 4. Pumping Oil Wells at Oil Springs, 1860s. *At first a separate steam engine supplied the power to pump each well, but as the number of wells increased and production began to fall off between 1863 and 1865, their relative scarcity in this region in terms of fuel and labour probably inspired a drive to a more practical power system.*

Plate 5. The Jerker System at Oil Springs, c. 1860s. *The solution to a more practical power system for pumping the petroleum wells of Ontario lay with subdividing power produced centrally and transmitting it to many points of application. This led to the introduction of the so-called "jerker system," in the 1860s. The advantage of this system lay in the cheapness of its operation over long distances when the power requirement was low.*

Although it is not certain where in North America this jerker system was originally introduced, it clearly was not a local invention. Rather, it represents a late stage in a long line of development of the *Stangenkunst* (roughly translated as "rod work with crank"), which had been a technique introduced into mining in European areas of German settlement in the mid-sixteenth century to transmit the power from waterwheels to pump mines located at a distant point.[85] Pennsylvania oilmen soon developed a more sophisticated and expensive version of the jerker system in which steel cables and iron rods replaced the wooden rods. This Pennsylvania system of pumping became internationally popular, but in the more marginal oilfields of southwestern Ontario the old jerker system with wooden rods persisted – even to this day.

Once pumped to the surface, petroleum had to be transported from the wells to the refineries, and the refined products to the markets. Gathering lines were laid from each well to a central tank; from there the petroleum was teamed, pumped, or sent by railway to refineries. The American petroleum industry had expanded so quickly that American refiners had worked out the technology for an efficient and inexpensive system of long-distance transportation by petroleum pipeline by the early 1870s.[86] The principle of transporting fluids by pipeline was not a new one, but the main problem was to find a weld for joints that would not leak. The operating costs of a pipeline were very low, and, unlike with railway transportation, they were relatively unchanging over time. The major costs consisted of securing the pipeline right-of-way and purchasing the pipe. An added advantage, and the reason for its popularity in the United States, was that it could break the control of producers and railways over the price and availability of crude. Elimination of this bottleneck meant that refiners and storage and tanking companies could become involved in any stage of the business. In contrast with the American petroleum industry the Canadian industry did not expand beyond the tiny, shallow-lying oilfields of southwestern Ontario until the twentieth century. Hence, long-distance pipelines were not constructed in Canada before 1890, although there was one short line built in 1880 to connect the Petrolia oilfield with a refinery at Sarnia. Similarly, beginning in 1880 oil could be transported overseas in custom-built oil tankers, but this mode of transporting oil would have had little impact on Ontario producers and refiners because their markets by that time were local ones.[87]

The salt resource, unlike petroleum, occurred in solid form and had to be dissolved before being pumped as brine. Since the technology for well-boring and for pumping the brine was similar to that for petroleum, little technological innovation was necessary. Production of brine, however, required a means whereby water could be injected into the rock salt deposit. This was accomplished by employing a double casing, surface water seeping or being forced down through the outer annulus and saturated water being pumped up through the core. The pressure of the water being pushed down to the salt measures aided in raising the brine so the pump had little lifting to do.[88]

Distilling, Refining, and Evaporating

Crude petroleum is a mixture of a great many chemical compounds, some of which are gaseous, some liquid, and some solid. All of these are mutually dissolved to form a liquid hydrocarbon. Crudes differ as to their characteristics. Some have a great deal of natural gas in solution, while others have very little; some are thick and heavy, and when refined give a large yield of lubricating oil; others may give a high percentage of illuminating oil, which constituted the most valuable output of refineries in the nineteenth century. Pennsylvania crude was an example of the latter, while Ontario crude fell within the former category.

Because illuminating oil and lubricants were the chief uses of petroleum before 1900, the process of distilling and refining crude therefore simply involved crude physical separation of the components without causing any real change in the chemical composition of the crude oil. Because the internal combustion engine and the petro-chemical industry lay in the future, so, too, did the processes of thermal conversion, catalytic conversion, and synthesis.

The process employed in the era of physical separation drew largely upon simple chemistry and the earlier technology used in refining kerosene.[89] Distilling the crude petroleum involved applying heat to small batch, horizontal cylindrical sheet metal "shell" stills or kettles – much like traditional medicine stills – containing crude petroleum, which was boiled and vapourized. The vapour condensed into components ranging from light fractions such as gasoline, naphtha, and benzine, through the illuminating oil, followed by intermediate and wool oils, the lubricating oils, to the tar residue which remained as an encrustation in the retort. Various grades of these products could be derived by controlling the processes of distillation, redistillation, and treatment of the products for improvement in colour and odour. As well, a variety of intermediate oils, tar by-products, and solids such as paraffin and vaseline could be produced. Before 1890, stills were relatively modest affairs with low capacities and not capable of making sharp distinctions in separating the various components. Historical evidence suggests that the typical refinery was capable of producing on average two thousand barrels per week and cost anywhere from $60,000 to $80,000 to construct and equip.[90] The bulk of the output of petroleum refineries was retailed in grocery stores.

Plate 6. Oil Refinery of the Producer's Oil Company, Petrolia, 1880s. *Petroleum refining before 1900 was characterized by crude physical separation in metal stills. Refiners chose between using horizontal or vertical stills, numerous types of heat applications, and several treating methods for quality. Between 1867 and 1873, pressure to increase the scale of operations led to the multiplication of stills with an increased production capacity and more extensive cracking capability.*

The principal innovations in distilling and refining in these formative years for the industry were in the area of refining illuminating oils. The earliest process of refining the illuminating portion in Pennsylvania and Ontario was to agitate the distillate with sulphuric acid to remove the tarry material, deodorize it, and improve its colour; caustic soda was used to carry this process further and also to neutralize the acid and the compounds the acid had formed. Afterwards, the oil was transferred to a bleacher, where it was exposed to the sun for a few days in order to lighten its colour, then barrelled. However, crudes with a sulphur content above 1 per cent were described as "sour," and as such they could not be deodorized with the standard refining process just described. The southwestern Ontario petroleum, unlike the Pennsylvania crude, was sour, and the smell from its high sulphur content threatened to ruin it for use as an illuminant. Individual refiners and chemists in this district struggled with the problem from 1868, when the lead process first came into use, until the 1880s.[91] The major breakthrough in North America to a commercially feasible process for refining and deodorizing sour petroleum was made in London, Ontario, by Herman Frasch, the celebrated German chemist. The Frasch method was developed between 1884 and 1889. Like those "improved" methods before it, it was based on the principle of a reaction between metallic oxide and the sulphur contained in crude mineral oils, but Frasch's went one step further and added copper oxide to the already saturated solution of crude oil and metallic oxide. Not only was the illuminating oil thereby rendered sweet, but also the copper oxide could be recovered for reuse.

Similarly, the individual characteristics of brine and the local circumstances in which it was discovered determined the specific process of recovering the salt. Solar evaporation was the oldest technique, but this method was not suited for all climates. The alternative method was to evaporate brine over open fires.[92] The English system employed iron pans with rectangular, sloping, or oblique sides arranged in rows in open buildings, called "blocks." Each block had a number of open fireplaces and flues for the distribution of heat. A second principal method was the Onondaga, named after a prominent New York State salt company that developed it around the year 1860; it also was referred to, from its locale, as the Syracuse method. This method employed cast-iron kettles or rectangular pans of iron or wood arranged and heated as in the English system in parallel rows, with one or two rows of sixty or more kettles to a block. The cost of a salt block ran about $3,000 ($1,500 for sixty kettles, $1,500 for the furnace).

Both the English and the Onondaga methods were adapted by the salt manufacturers in the western peninsula district, who were also innovative when it came to improving the method for heating brines. As early as 1842 the English were experimenting with heating brines by steam coils.[93] Heating liquids directly by means of open fire located under vessels was dirty, dangerous, and more important to securing profits, inefficient, hence costly. Salt manufacturers, as well as petroleum distillers and refiners, adapted innovations with the use of steam enclosed in and circulated through coils placed within vessels or in steam chambers to replace the old methods of heating with open fires.[94] The advantages of the new ways were faster heating, greater safety, and increased efficiency. Moreover, the intensified heating method meant that the vessels themselves could be constructed with less metal or, in the case of brine evaporation, even entirely of wood, resulting in even greater savings to manufacturers. Salt manufacturers quickly recognized the advantages of steam heating, which became standard practice throughout the Ontario salt district in the 1870s.

In summary, it should be clear from this overview that a genuine transformation of mining technology occurred during the fifty-year period covered in this study and that the transformation involved entirely new systems of mining and not merely improved machinery and larger plants. In contrast with the earlier period, when European mining regions were at once the innovators and the sole beneficiaries of their innovations, now innovation and diffusion occurred on a more global scale, as mines were being brought into production all over the world.

3

THE DIFFUSION MECHANISMS

Technological change in mining on the nineteenth-century Ontario frontier mainly resulted from adopting and adapting such new techniques as those just described. Once new technologies became available, they reached the Ontario mining districts in a variety of ways.

First, there were the traditional routes. Machinery and equipment which embodied new techniques were imported, and techniques were carried as skills by mine managers and workers who were brought to Ontario.[1] The importation of both capital goods and skilled personnel was closely related to the high rate of direct foreign investments. Not surprisingly, foreign owners usually introduced and applied mining systems – including specialized machinery and skilled workers – which were most familiar to their mine managers and agents.

The diffusion of new mining techniques also depended upon the newer medium of publications. Inventors, or their agents, and manufacturers promoted the various machinery and methods by means of trade catalogues and even by personal visits to mine sites. However, this was not a well-organized activity anywhere before 1890. Of far greater importance to the rapid diffusion of technological changes in mining was the work of professional associations and of governments.

During the second half of the nineteenth century, mining engineering began to develop as a separate profession, and professional associations and governments began to take responsibility for technical and engineering education.[2] International networks of formally trained, reform-minded mining engineers, geologists, and mineralogists were spreading information on technological advances through their organizations, which published journals, conference proceedings, and special reports. By the same token, by the final quarter of the century it had become common for governments to publish routine and special reports on mining, which often included highly descriptive and statistical accounts of technological changes.[3] All these developments within the wider world of mining were to be crucial in the rapid dissemination of new techniques and machinery in the Ontario mining districts.

As well as being a time when the new associations and national governments began to publicize new mining techniques, there was also tremendous growth in the number and distribution of privately published books and serials devoted to applied science and technology.[4] Moreover, the innumerable trade fairs and especially the hundred or so international exhibitions held at various times and in many countries accelerated the development of industry and technological advance generally. They also provided important opportunities for specialists in a small, developing country like Canada to participate in the exchange of ideas and information on mining developments and new technology and, more importantly, for Canada to attract foreign investments.[5] Incidentally, the responsibility for organizing Canada's early exhibits rested largely with the Geological Survey of Canada staff, the same men who were frequently contacting and advising provincial mine operators.[6]

About the role of people and publications in introducing improved mining technology to Ontario there is, unfortunately, only indirect evidence. Such evidence as exists will be discussed here in a general fashion, leaving the details and specific individuals' contributions to technological diffusion to be treated in the chapters on the three mining districts.

PEOPLE

The international communications network in the mining industry consisted first of interested specialists: experienced miners and mine managers, inventors and machinery promoters, and professional engineers and public servants, all of whom were agents of technological change in the pioneer period of mining in Ontario. Because of their knowledge of successful techniques gained through practical experience and professional training and membership in various associations or mining institutes, they were able to keep up with new developments and were in a position to spread information during their movement about the province from advanced to less-advanced or declining to newly opened mining districts.[7]

Since extensive mining operations were going on in Europe, mining methods and mining men from there were available to the early North American camps.[8] Indeed, Cornish miners were to be found in mining camps throughout the world, especially once the Cornish tin industry entered a permanent decline in the 1860s.[9] As the Michigan copper region was opened to full development in the same decade, increasing numbers of Cornishmen and others gained experience there. Many of the early mine managers in North American camps were former Cornish miners who had won promotion through their experience, skills, and ability to handle men.

Naturally, only the most productive mining regions actually wanted or drew highly skilled miners. In the case of Ontario, fewer skilled miners would have been attracted to the southeastern district than to the Upper Great Lakes, for example, while the western petroleum and salt district literally had to generate its own skilled workforce. In any case, little record survives from which to identify, let alone to trace the lives of, the individual miners who worked in the mining districts of Old Ontario.

The extent to which inventors and machinery promoters influenced technological change in Ontario mining is likewise difficult to ascertain from existing records. Patented inventions were marketed through personal contact with mine operators, through periodicals, and at exhibitions. Machinery salesmen and patent agents needed to establish a reputation for their equipment by collecting testimonials from mining men, advertising, and, most significantly of all, gaining a successful trial of their equipment in a reputable mining operation. Being a large-scale borrower of technology, the territory should have attracted the attention of such men. However, since most Ontario mining operations in this period were so marginal and ultimately shortlived, they really offered little incentive for such trials. Only the few promising mining operations, such as Silver Islet and the Duncan Mines in the Thunder Bay area and the Dean and Williams Mine in the Marmora goldfield, proved exceptions.

The painstaking campaign a Windsor patent agency undertook in the Marmora gold region to market a new American steamdriven stamp battery is an ideal case in point. In 1870 and 1871, the agency, Macdonell and Foster, attempted to promote the previously-mentioned quartz crusher invented by an Indiana mechanic, I.W. Forbes, of which they owned a three-quarters interest and the Canadian rights. This trial was unsuccessful, but their approach to promoting the invention as it is detailed in company records provides a rare insight into this sort of activity.

By the time Macdonell and Foster arranged for a trial of the Forbes patented stamps in 1870, its promotion had already been delayed for a year while radical changes were made to overcome initial faults in design.[10] After that, the inventor managed to collect testimonials from two Chicago machinists and installed his batteries at two mining operations in Utah.[11] At this point, Macdonell and Foster began their assault on the Canadian market. Contacting the chief administrator of the Nova Scotia mines for a list of provincial mine operators, they sent them promotional literature on the batteries.[12] They built their hopes around a successful trial in the Marmora goldfield of the southeastern district, hoping that the published reports plus a visit from the chief administrator of the Nova Scotia mines would lift the invention from the experimental to the commercial stage. Arrangements financially favourable to the operators were made in strictest confidence with the area's most prominent gold-mining operation – the Dean and Williams Mine.[13] The mine operators agreed to telegraph a Toronto newspaper with news of the arrival of the battery, and Foster arranged to purchase fifty copies of the issue for use in future publicity ventures.[14]

When the battery proved unsuccessful, Foster was anxious to move it to a nearby site for a second try since he feared an admission of failure would damage their business interests.[15] Foster also urged the inventor to make adjustments to the design while he continued to work on obtaining a second trial for the machine. It was clear to Foster that the supposed advantages of this equipment would have to be proven for, as he wrote to his partner: "The only question that people care to have answered at the outset is 'what has it done?' If you cannot answer that satisfactorily you will only injure its cause by advertising it in any other way."[16] The agency turned its attention to the newly opened silver field in the Upper Great Lakes district. A Detroit director of the Silver Islet mining operation, E.B. Ward, apparently expressed interest in the battery, but only if it could be shown capable of doing what was claimed for it. The agency offered to arrange for a trial in Toronto, but given the unimproved state of the invention and the inventor's indifference to establishing a Canadian market for it, the company seems to have dropped the project.[17]

This experience points to the inefficiency of loose, informal arrangements for the promotion and sale of inventions. This is in part because of the time-lag between when an invention was conceived and when it became practical. A detailed study of the invention and promotion of Sperry's Coal Puncher by an American coal operator in 1889 reveals a common procedure for testing, publicizing, and marketing essentially unproven inventions.[18] Because the machine was produced by an operator, it received immediate trial in the field. After field tests proved successful, the Goodman Manufacturing Company was incorporated. No catalogue was issued. For publicity, the company relied instead on the technical press, then simply purchased a stock of copies for distribution to potential customers. All the while, no machines were actually operating at the mine. Underlying these observations concerning the Forbes Quartz Crusher and Sperry's Coal Puncher is the suggestion that only large, well-financed companies, like Fraser and Chalmers, could really break into the mining equipment industry. They did so by buying up all relevant patents and establishing an international sales network.

To return to the main theme, those persons who were advising on the adopting of technological innovations were increasingly apt to have been formally trained and formally connected to the international communications network on mining. Before 1870 most mining engineers in North American camps had received their training in European schools – the London School of

Mines, Ecole des Mines (Paris), and the oldest and most prestigious of all, the Royal Mining School of Freiberg.[19] The Geological Survey of Canada and the United States Bureau of Mines both called upon their respective governments to establish a national school of mines along lines similar to the European schools but with more practical aspects.[20] However, the sheer vastness of the two countries and the widely differing conditions for mining in various districts alone made such an undertaking impossible.

Despite these failures, formal education of mining engineers in North America did begin as early as 1867 with the opening of the Columbia School of Mines, followed in a year by the Massachusetts Institute of Technology. By 1892, sixteen institutions in the United States were graduating mining engineers, though it is not known how many, if any, Canadians attended them.[21] A total of 871 graduates in mining engineering were produced between 1867 and 1892, of which Columbia School of Mines was responsible for roughly half, 402, and the Massachusetts Institute of Technology, another 126. As part of their training, students in these programmes were exposed to the standard and latest milling equipment of the day.

In Canada, McGill University offered the first Canadian degree for mining engineers – the School of Applied Science, established in 1871-72, offered majors in civil and mechanical engineering, metallurgy and the exploitation of mines, and practical chemistry.[22] Twenty students had graduated in mining engineering by 1892. Beginning in 1873, formal instruction in mining was also available at the School of Practical Science, University of Toronto, though a degree in mining engineering was not possible there until 1892.[23] The school's annual reports to the Ontario legislature for the first few years reveal that the emphasis was placed on the mechanical and manufacturing arts, rather than on mining. The school at least did establish an extensive library in connection with technology and mining, a laboratory for examining mineral samples from provincial mining districts and mines (though no mining equipment was provided), and a geological museum.[24]

It is unlikely that the effects of these Canadian programmes were felt in Canada for a few decades. For one thing, the numbers of graduates were tiny, especially during the first few decades. McGill, with the earliest and largest programme in mining engineering, averaged only roughly one graduate per year in the period 1873 to 1892.[25] Also, according to McGill's graduate records, most of the graduates in mining engineering found employment outside the country, in the American West, Mexico, and South and Central America.[26] What of the mining engineers who were employed in Ontario? They had been educated in Great Britain, Saxony, and France, not in Canada or the United States, as the following brief biographies reveal.

Among the first of provincial mining men with broad experience and knowledge of new technology, perhaps none had as long-term impact on early mining in Ontario as E.B. Borron. Borron was scientifically trained, experienced in both the Scottish and Michigan mining fields, and for a few critical years during the 1850s copper boom, he supervised the famous Bruce Mines on Lake Huron, where he experimented with numerous British and North American techniques for dressing and smelting the copper ores.[27] In 1865 he patented a two-hundred-acre mineral location south of Mississagi River on Lake Huron.[28] He continued to have an impact on mining in the area after 1867, when the newly constituted province of Ontario appointed him land agent and inspector of the Sault Ste. Marie Mining Division. Later he worked in the district as the first stipendiary magistrate for Northern Nipissing, an area that included the country north of the Upper Great Lakes and west of Lake Timiskaming, carrying his experiences in mining operations and technology throughout the newly opened mining territory.[29] Moreover, he was in a position to influence the province on resource matters – which he did. He was a principal witness before the Ontario Royal Commission on Mineral Resources in 1888, and in his testimony he pressed the Ontario government to create a school of mines.[30]

Thomas Macfarlane carried on a similarly distinguished mining career in Canada beginning in the 1860s. Macfarlane was a British professional mining engineer and the former manager of a Norwegian copper mine, and he is important for his vast experience with mining developments in Ontario and surrounding mining districts. He served the Geological Survey temporarily in 1865, yet in that single season he investigated both the east shore of Lake Superior and the north Hastings County area of the southeastern district.[31] Afterwards, he worked as a mining engineer for Canadian and American interests in Canada and the United States. He led the exploration party for Montreal Mining Company interests in the Lake Superior region, for example, and in 1868 he was in charge of the party that discovered the famous Silver Islet strike. For a time he managed the Silver Islet Company's Wyandotte Silver Smelting Works, near Detroit, where a number of innovations were successfully combined to produce an efficient new process.[32] He also managed the world-famous Acton Copper Mine in the Eastern Townships of Quebec, and he patented an aspect of a new process for smelting local copper ores.[33] Finally, Macfarlane presented influential testimony at the second Select Committee (1884) hearings on the Geological Survey, urging the Survey to concentrate their efforts on mineral-rich districts; provide descriptions, analyses, and statistical reports on mining activities that would serve the interests of potential investors; and employ more mining specialists. These recommendations were acted upon by the committee.

The work of three other men, E.J. Chapman, W.H. Merritt, and Edward Peters, Jr., represents the sort of independent geological and mineralogical work being conducted by private individuals in the province. Chapman, a University of Toronto professor who had studied mining in Germany and engineering in England, was strongly connected with the work of the Geological Survey and with private mining interests.[34] He made assays and gave general advice as to the probable value of prospects, becoming a specialist on the southeastern district in the 1870s. He became professor of mineralogy at the Toronto School of Practical Science shortly after it started up in 1873, and in that position he taught students mineral identification, the use of the blowpipe, and assaying techniques – all of which were requisite skills for the judicious application of mining technology. The department examined mineral samples free of charge for explorers and farmers. He patented several inventions of his own on a new chemical treatment and by-product recovery process for the gold quartz of the southeastern district, and he provided important testimony to both the Select Committee hearings on the Geological Survey and to the Ontario Royal Commission on Mineral Resources – in all instances pushing for formal training of mining men.[35] W.H. Merritt was the grandson of the prominent pioneer entrepreneur of the same name and a graduate of the London School of Mines.[36] Merritt was active in eastern Ontario in the 1880s as a mining consultant and served as a member of the Ontario Royal Commission on Mineral Resources. When the provincial School of Mines opened its doors at Queen's University, Kingston, in 1893, Merritt was on hand as the first lecturer in mining engineering. Edward Peters, a mining engineer, had been educated as a metallurgist in Freiberg. He worked for a time in Colorado as a government assayer and in mining in Portugal, Hungary, and Michigan before he joined the Canada Copper Company in the late 1880s as the ingenious manager of its copper-nickel operation in Sudbury.[37]

The career public servants, however, perhaps were the most instrumental group in spreading the word about new mining techniques and apparatus. Not only were they scientifically trained and experienced, with access to the latest developments as promulgated in professional associations and technical and professional literature, but also, because of the time they spent in the field, they also were "men on the spot." As such, they were in a favourable position to develop expertise on the special nature and modes of occurrence and hence on the most appropriate technology for developing, provincial mineral resources. Of all the public servants active in promoting mineral development and the spread of technological information, none were so numerous and pervasive as the permanent members of the government-sponsored Geological Survey, who were noteworthy for the long periods of their individual service and for their field work in special territories. All of them to some degree interested themselves in the practical aspects of mining and in the judicious use of new machinery and techniques.

Three waves of Geological Survey staff corresponded to the periods of expansion and the shifts in emphasis from theoretical work towards practical mining matters. The backgrounds of the most important members of the

Survey Staff working in Ontario indicate the way the geological profession changed over the decades from professionals with primary interests in land surveying and geology to men formally trained and deeply interested in the practical aspects of mineral development.

In the pioneer group is William Logan, founder of the Survey in 1843, director until 1869, and its most influential member. A Canadian by birth, Logan was educated in England, and in 1831 he was made manager of his uncle's copper mine in Swansea, Wales.[38] There he became involved in professional mining and geological associations and was thoroughly trained in the latest geological principles by the time he returned to Canada. His own field work determining the general geological profile of the province involved unlocking the mysteries of the Precambrian Shield and offering practical technological advice in the first serious mining developments in the Upper Great Lakes district.[39] His first and long-time assistant, Alexander Murray, made Canada West districts his specialty during twenty-two years with the Survey. Murray left in 1864 to become director of the Newfoundland Geological Survey. The third of the Survey pioneers working on Ontario was T.S. Hunt, who served as Survey chemist and metallurgist from 1848 to 1872. His chemical investigations were supplemented with extensive field surveys of practical use to mining interests. Notable in this regard were his work in the petroleum and salt region of the western peninsula district, his innovative theories on petroleum, and his commitment to collecting and publishing the logs of local well-boring activity at an early and critical stage of salt and petroleum developments.[40]

A second, younger wave of staff prosecuted Survey work in the 1860s and 1870s. These new men, who were more inclined to emphasize economic geology in their field work and reports, included the likes of Robert Bell and Henry G. Vennor. Bell, whose father had been the leading amateur geologist in Canada and a friend of William Logan, was a scientist whose specialty became the northern district of Ontario. He also taught chemistry and natural science at Queen's University, became a major shareholder in the Frontenac Lead Mine, and served as one of the five commissioners appointed to the Ontario Royal Commission on Mineral Resources.[41] The other Survey member to focus on Ontario mining districts in this period was Vennor, whose specialty was the southeastern district where he served exclusively from his first year with the Survey, 1866, until he left in 1881. His work was particularly helpful to contemporary mining interests because he concentrated on mineable deposits. In his reports he noted prospects and provided highly informative mining studies of the entire district.

The third wave of Survey staff reflected the increasingly practical directions and future trends of the Survey on behalf of mining interests. It included several mining engineers and statisticians hired in response to the Select Committee recommendations of 1884. In addition to collecting and analysing mineral and mining statistics, they launched a programme of field work as a basis for historical monographs on specific mineral-rich regions (as Macfarlane had recommended). One of these men, Eugène Coste, a graduate in mining engineering from the Ecole des Mines and the Royal School of Mines in London, worked for the Survey only a few years, from 1883 to 1889, when he became involved in pioneer commercial developments in natural gas both in the western peninsula district and in the North West Territories.[42] The other mining engineer made an even more impressive contribution. Like Coste, E.D. Ingall was a graduate of the Royal School of Mines, and his career with the Survey extended from 1885 right up to 1928.[43] The work of their assistant, H.P.H. Brumell, focused on mining statistics, but he was also involved in investigating natural gas and petroleum operations in southwestern Ontario.

While it is difficult to assess the impact of word-of-mouth on technological change in pioneer Ontario mining operations, it is safe to assume that the role of people in the transmission of techniques was significant, especially in the initial stages of mining development and in the spread of technology within a specific region. As long as mining operations relied on strictly traditional techniques and familiar minerals, publications were not likely to play as considerable a role in the diffusion of mining technology as people. But with large-scale development and the economic application of machine power and science to mining operations after 1870, the demand for published materials on mining and new mining techniques naturally increased.

PUBLICATIONS

This aspect of technological diffusion is especially important to study because publications were tangible, educative, and crucial to the *rapid* dissemination of reports of technological innovations being developed around the world after 1870. A simple enumeration of publications available to the Ontario mining industry does not prove that all or any of them were read or were influential in the territory. It is unlikely that practical miners, for example, would have subscribed to the professional literature on mining. However, a great deal of richly detailed popular literature on science and technology, such as the *Mechanics' Magazine* and *Scientific American,* were available to them. Publishers on engineering, science, and technology borrowed extensively from the professional literature and from one another, so their publications became compendia of reports on the latest techniques. This was a particularly common practice for illustrated journals and magazines such as *Scientific American.* Local newspapers also picked up these stories and carried them, along with news of the latest mineral discoveries, rumours of interest on the part of capitalists, information regarding the comings and goings of mining men, advertisements for mining equipment, announcements of the arrival of new machinery and equipment bound for the mines, and much more besides. Hence, a reader could keep up-to-date with many developments in mining technology by quite casual reading.

By an examination of what was available, in what form, and when, it is possible at least to speculate about the general role of publications in the spread of advances in mining techniques. The contemporary international English-language literature that carried information on changes in mining technology was immense.[44] There were plenty of monographs on mining, but serials were even more prevalent, and they had the advantage of reporting and publicizing the newest practices and equipment. Because the mining industries of various countries were at differing stages of advancement and organization, news of mining developments, including new methods and machinery, was broadcast internationally by a broad range of journals and magazines, and reports on technological innovations in mining were found in a wide variety of contexts. Currently available issues of 41 major non-governmental publications from Great Britain (20), the United States (13), and Canada (8), selected from a long list of those possibly relevant in the published guides, have been examined to discover the nature, origins, and timing of machinery and methods available to provincial mine operators (see Bibliography).[45] They include professional publications of mining, mechanical, civil, electrical, and other engineering associations (3); statistical, philosophical, and geographical associations (3); mineralogical societies (1); and natural science associations (1). The more popular periodicals include the typically combined artistic, literary, and historical journals (2); natural history magazines (1); general mining magazines (6); other magazines, on coal-mining (1), engineering (2), and patents and inventions (4); and magazines directed at mechanics (1), and architects and builders (1). A very disparate collection of publications, to be sure; yet each one contains information on mining techniques.

A few general observations are in order. First, a quantitative and qualitative increase occurred by 1870 in the publications carrying news of mining and associated technological innovations, and, in particular, a growing number of publications were devoted exclusively to mining. Second, and even more significant for the diffusion of information to Ontario, was the strongly evident shift by 1870 of the centre of publishing on mining from Great Britain and Europe to the United States.

Not until 1882, when publication of the *Canadian Mining Review* began, was there a periodical published in Canada that was devoted exclusively to mining. The *Review* was a monthly devoted to promoting the mining interests of the entire Dominion, but it gave special attention in its first years to the mining interests of the Ottawa Valley and to mining in the disputed territory that today comprises northwestern Ontario.[46] Only with the establishment of the Ontario

Bureau of Mines in 1891 did reports on Ontario mining began to be regularly published. Prior to that time, international news of mining and of new techniques – including some accounts of developments in Ontario – appeared occasionally in the journals of Canadian literary and scientific societies and in popular scientific magazines.[47] One of the earliest such Canadian publications to carry news of mining techniques was the *Canadian Journal of Science, Literature and History,* sponsored by the Canadian Institute, which began in 1852. Along with the *Canadian Naturalist and Quarterly Journal of Science* (1856) and the *British American Magazine* (1863), it routinely contained articles on the work of the Geological Survey in mining districts. The *Canadian Patent Office Record* (1873) was the first popular Canadian magazine on Canadian inventions and technological change; it was a valuable information source for Canadian interests because it published abstracts of patented inventions and reprinted articles on mining and mining technology culled from a wide range of foreign publications, notably the *Scientific American.* The two professional science and engineering periodicals that devoted considerable space to developments in mining engineering and technology started up only in the decade of the 1880s, when the *Canadian Mining Review* (1882) also began publishing. These were the *Proceedings and Transactions of the Royal Society of Canada* (1882) and the *Transactions of the Engineering Institute of Canada* (1887).

This reliance on external publications for news of developments in mining and mining technology might have been a serious constraint on the introduction and diffusion of innovations in mining technology in Ontario had it not been for the very presence of the Geological Survey of Canada, which, as indicated, made a singularly important publishing contribution on domestic and even foreign developments. It accomplished much through reports of its members' field work, research and analyses, and a proliferation of descriptive and analytical articles by its officers, which appeared in Canadian and foreign literature and in their own regular publication. They were based on Canadian conditions and experience and so were especially useful for Ontario mining operations. Annual reports from the director of the Survey began to appear in 1845, but they were somewhat irregular until Confederation.[48] After Confederation, responsibility for the Geological Survey passed to the dominion government, whose first report from the Survey containing reports from the years 1866 to 1869 was published in 1870. Under the authority of the dominion government, the yearly reports of the Survey continued on a fairly regular basis.

The earliest reports outlined the principal geological features of the Province of Canada and were invaluable for future mineral developments generally. In the next fifteen years the entire territory of Canada West was surveyed, with special attention to known mineral regions. Of particular importance for the first mining operations, early attention was directed to the northern division, the Precambrian Shield territory, where reports of copper-ore discoveries in

Upper Michigan in the early 1840s led to a rush.

During the decade of the 1850s, the Geological Survey continued to conduct extensive surveys of Canada West, which had been wholly investigated at least once by the late 1850s. The Survey continued to publicize its findings domestically and abroad in impressive displays and publications at the first two great international industrial exhibitions, at London in 1851 and at Paris in 1855.[49] As mentioned, participation at these and subsequent exhibitions placed foreign capitalists in touch with investment possibilities in Canadian mining and also put the Survey staff and others representing Canada in contact with the latest mining methods and machinery. This opportunity to augment their knowledge of new technology was significant because of the amount of time they would be spending in the mining regions.

The end of this first geological descriptive phase was accompanied by a call from entrepreneurs and others for more information on mineral prospects and mining activities, including techniques employed, within the province. This became the subject of an early investigation by a Select Committee appointed in 1854 to look into the best means of making Survey findings available.[50] They recommended that the publications of the Survey (revised reports and fuller future reports) should be printed in larger editions – an impressive ten to twenty thousand for *Annual Reports,* and two to three thousand of the plates and illustrations – and that a museum and library be established. Diffusion of information on Canadian mining matters, including technology employed, was to be generous. Disposition of Survey publications was to be as follows: four copies to every member of the legislature; a copy to each university and college, to literary and philosophical societies; to public libraries and mechanics' institutes; to grammar, model, normal, and common schools; and to municipalities with a school or library. In part, this was a bid to encourage a taste for geological studies on the part of amateurs. Free copies were also to go to the governments of British colonies for distribution and to each of the principal libraries and learned societies in the United States and Europe. Lastly, some copies should be available for public sale at cost price.

While no complete record of the entire disposition of the Survey's reports survives, they are listed as exchange items in the journals of several societies and magazines, as, for example, *American Journal of Agriculture and Science; British American Magazine* (Toronto); *Canadian Journal of Science, Literature and History; Proceedings of the Royal Geological Society* (London); *Transactions of the Canadian Society of Civil Engineers;* and *American Geologist.* Moreover, beginning in 1873 lists of publications received and exchanged are given in the *Annual Report* of the Geological Survey, which give some notion of Canada's involvement in the widespread international network for the transmission of scientific and technological knowledge in mining. This routine exchange of technical reports meant that Geological Survey staff had the same opportunities as members of any professional society to keep up with international developments. Members of professional societies typically kept up-to-date in their field through establishing a well-stocked library. When the Canadian Society of Civil Engineers, for example, formed a library in 1887, they immediately secured exchanges with twenty-nine engineering and kindred societies and then began publishing lists of their library holdings.[51] The significance of the wide distribution of Geological Survey publications and the growth of the Survey library for a pioneer economy like that of Canada was that they made staff members and the public better informed of the latest experiments with, and applications of, mining technology.

The Survey's new emphasis in the decade of the 1860s on the more practical aspects for mineral development and technological advance is reflected in its publications. *The Geology of Canada,* published in 1863, marked this turning point and was an impressive scholarly contribution to international scientific literature, while at the same time being an invaluable reference work for capitalists and practical mining men. Of especial importance, it provided the first details on the geology of the Canadian Shield, the region that would prove to be Canada's greatest source of mineral wealth. When the geological map and atlas of Canada appeared in 1866, in time for the second Paris exhibition, they significantly reduced the amount of guesswork associated with mineral exploration in Ontario. With the basic primary geological task completed by the time of Confederation, the Survey was able to follow the major Ontario mineral strikes of the next two decades from their beginnings. At the same time as the staff continued this work, it was also extending its operations to the examination of all sections of the former Rupert's Land and other newly acquired territories.

In line with their new emphasis on the practical aspects of geology and with trends in other countries, the Survey attempted to institute the systematic collection of records of mines and statistics of their production and levels of operation, including details of technologies employed. Charles Robb headed up the statistical survey, which began in 1870 when mining in Canada appeared to be coming into its own. The enterprise was anything but an immediate success. Of the several hundred mining operations in Ontario, only a dozen responded to Robb's appeal, and even then, the responses were woefully incomplete. The early efforts at collecting and publishing mining statistics in Canada and of monitoring productivity and technological advance failed for two important reasons: the natural reluctance of mineowners to reveal the true nature of their operations and the organizational problems for the tiny Geological Survey staff in simply collecting and compiling such a record.[52] While the first of these could not be easily overcome, the task of administering a programme of mineral and mining statistics could be helped by appropriate legislation.

Following the appointment of a second Select Committee to enquire into work of the Geological Survey in 1884, the resulting new Survey Act, passed the same year, instructed the Survey to collect and publish annually complete statistics of mineral production and of the mining and metallurgical industry, to collect and preserve available records of wells, mines, and mining works, and, of even greater significance for the diffusion of technology, to produce detailed economic mineral surveys and reports.[53] As a further indication of their intention to encourage the practical work of the Survey, the Select Committee also instructed it to appoint a mining engineer with the rank of assistant director. Within four years the statistical report of production, exports, and imports of minerals and fuels, based on the results of three thousand returned questionnaires, was appearing.[54] While the original returns have been preserved in the Public Archives of Canada and contain important details relevant to mining technology, there are fewer than fifty returns for the entire mining industry of Ontario, and these are rather incomplete.[55]

These new concerns of government with the practical aspects of mineral development were nowhere more evident than in the work of the Ontario Royal Commission on the Mineral Resources of Ontario, convened in 1888. This was the second of four provincial royal commissions on resources appointed between 1880 and 1892, and like the others – on agriculture, fish and game, and forest reserves – it was expected to elicit information contributing to the objective of asserting greater provincial control over colonization and systematic resource development in the province. More immediately, too, the province was interested in starting to develop the resources of the vast northern territory. These were opened to development by construction of the transcontinental railway section across the disputed territory which had been originally awarded to Manitoba and then to Ontario, but had only recently been confirmed to the latter province. Most of the commissioners were men with experience with Ontario mining – Robert Bell, who by this time was assistant director of the Geological Survey; William Coe, a Madoc explorer and miner who was developing the north Hastings iron fields at Coe Hill; William H. Merritt, the Toronto mining engineer; and Archibald Blue, deputy minister of agriculture, secretary of the Bureau of Industries for Ontario, and future deputy minister and head of the Ontario Bureau of Mines. The commissioners travelled extensively throughout the province, visiting mine sites and examining close to two hundred witnesses. They also travelled outside the province to observe foreign mining districts and schools of mining. In the wake of its lengthy report in 1890, a provincial bureau of mines and school of mining were established. Thereafter, the primary responsibility for development of provincial mineral resources effectively shifted from the dominion agencies – notably the Geological Survey – to the provincial government's Bureau of Mines. The extension of the northern boundary of the province and the establishment of the Ontario Bureau of Mines spelled the end of the pioneer era of mining in Ontario.

This summary shows something of the experimentation with, and improvements to, technological innovations applicable to mining during this half-

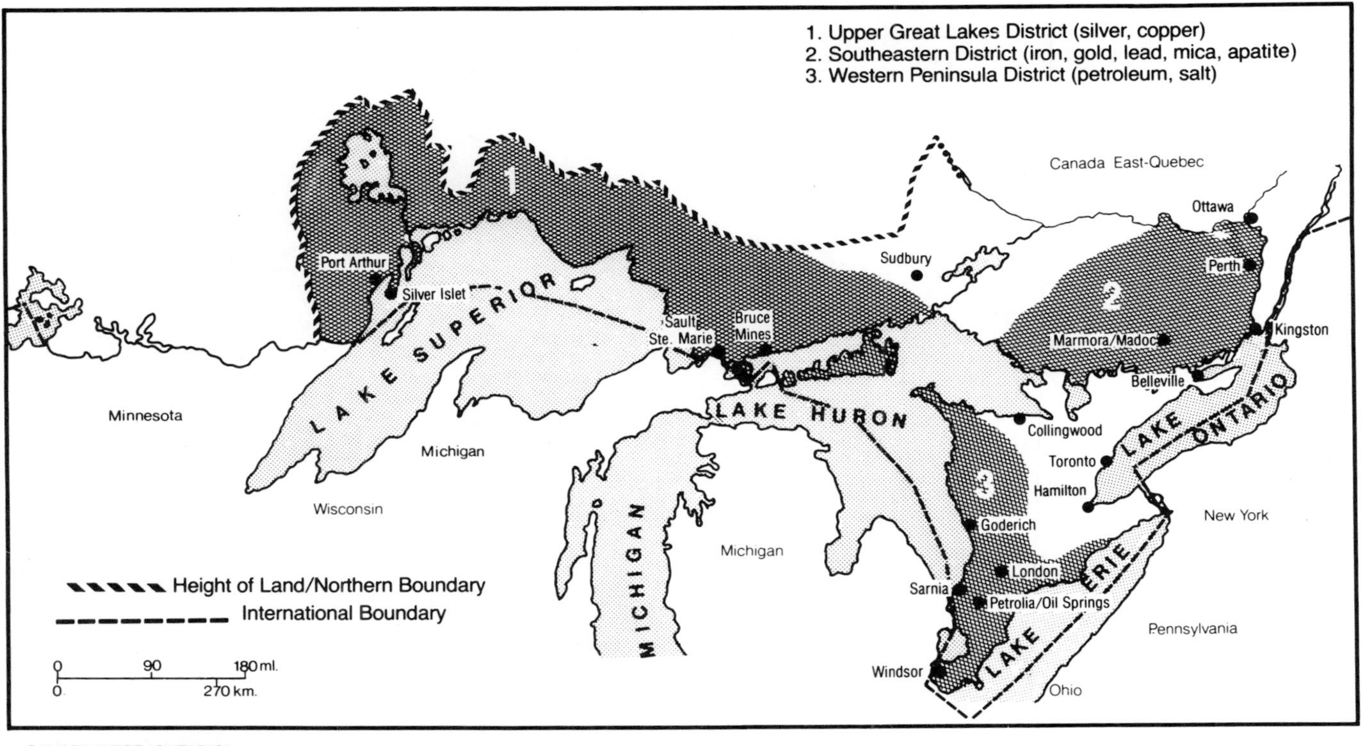

MAP 2. The Three Mining Districts of Canada West-Ontario, 1841-1891.

century of revolution in mining technology. Despite the absence of a provincial bureau of mines before 1890, provincial mine operators had direct access to information on many of these innovations. The information arrived with knowledgeable individuals, many of whom were scientifically trained public servants engaged in field investigations and inspections of the province's mining districts, and with numerous widely circulated periodicals. Thus, Ontario mine operators were in a position to know about innovations being developed outside the province. This was especially true because many of those innovations were being generated in nearby United States' mining districts, where the geological setting of mineral occurrences and the physical conditions were similar. But were Ontario mine operators in a position to apply the new technology?

The three case study chapters that follow examine the varying difficulties facing mining operations in the particular districts. They document the technological solutions adopted following the advance from exploration and testing to full development, then the inevitable depletion of the richest, most accessible deposits and the falling off in profits. With the depletion of the richest deposits, mine operators had to choose between working less accessible rich deposits farther afield from markets or located at greater depth in the existing mine or else undertaking innovations that would make it more economical to export and market leaner or more complex materials. Following the attempts to reduce the costs of production and transfer, came decline, perhaps a series of rejuvenations in response to mineral price rises or technological advances, then eventual abandonment. In looking at the introduction of technology, these chapters identify the leader operations. Last, they assess the relative weights of engineering, transportation, and economic factors and of such important internal forces as fuel requirements in determining the implications of the adopted changes on mining operations themselves, and for the general development of the regions in which they were situated.

4

UPPER GREAT LAKES DISTRICT

For half a century the mining regions of Lakes Superior and Huron appeared to be North America's answer to Cornwall. Of the three Ontario districts in which mining took place, this, the Upper Great Lakes, was the only one that government viewed as a true mining frontier. The Province of Canada-Ontario governments improvised policies to regulate mining in the Upper Great Lakes in response to changing economic circumstances, but, in general, they were preoccupied with settling the agricultural frontier, so they left the initiatives largely in the hands of the private sector.[1]

The Upper Great Lakes mining district is a rocky, heavily forested, thin belt of the Shield along the north shores of Lakes Superior and Huron. The principal mining region until the 1880s extended from the international boundary at Pigeon River a thousand miles eastward to La Cloche on Lake Huron, including nearby islands. This belt of land is fairly homogeneous both geographically and geologically, containing irregular occurrences of precious and base metals, though in this period only copper, gold, and silver were mined.

Unfortunately, there were formidable physical constraints hampering the discovery of worthwhile mining locations and limiting the scale of subsequent operations. The former fur-trading territory of the northern Shield was very much a wilderness outpost. Since there was no arable land local sources of labour and markets were non-existent, and each mine had to be entirely self-sufficient. Also, there was no handy nearby source of coal to fuel such essential mining equipment as pumps, hoists, and ore concentrators. Timber on the locations could serve as fuel, but wood is not as efficient as coal, and in any case construction and mine-timbering needs came first. The dense covering of forest and brush disguised mineral outcroppings, making them more difficult

Plate 7. Silver Islet Mine (1870-1884). *Silver Islet was the one Ontario mine to achieve a fame in its day that ranked with the best silver mines of the American West. It was here that the major technological advances and mining strategies of the day received their regional trials. Some of the company's records have been preserved, which is a rare occurance for such early mining ventures.*

to locate. Furthermore, the veins were irregular both in size and course, and the metal content of the ores found at any depth usually proved both low and uneven in content. For half the year ice and snow prevented development work and all movement of freight into and out of the district. Mine management, therefore, found itself faced with the difficult choice of laying off its workers, knowing that they likely would not return again in the spring, or investing heavily in a winter's supply of fuel and provisions to sustain a largely idle workforce. For these reasons of labour scarcity, geology, and climate, rich veins of mineral deposits nearest the surface and along the shoreline tended to be tested during the summer months; then operations here usually ceased altogether until a time when more extensive development, especially the installation of pumps, and a permanent workforce was required.

Transportation within the region was also uncertain and costly, depending upon canoes, sailboats, or privately owned steamers belonging to the fur-trading companies, and water travel was restricted to only seven months of the year. Moreover, it was especially hazardous owing to treacherous storms and to the ironbound coasts that played havoc with ships' compasses. Then there were the rapids at Sault Ste. Marie, which interrupted navigation between Lakes Huron and Superior.[2]

The only supply centres in the early period were the old Hudson's Bay Company fur-trading posts of Fort William and Sault Ste. Marie, both of which were poorly equipped for mining purposes. Since access to the bush-covered interior was limited to boat and canoe, the mines tended to be strung out along the coastline and islands. The distribution of locations also was uneven, since most serious exploration was concentrated on the vicinities of the two supply bases.

The completion of the Sault Ste. Marie Canal in 1855 had a dramatic impact on the copper-mining developments on the American side of Lake Superior, but alone it was not sufficient to promote activity on the Canadian shores, where minerals were not found to be as plentiful as in Michigan. Only the Bruce Mines, located on Lake Huron, were actively developed before 1870. Even with the discovery of a bonanza at the Silver Islet Mine in 1870, the Lake Superior section of the district remained a remote outpost. The deep levels at which the silver occurred made necessary a great deal of capital equipment and large, regular supplies of coal – both of which had to be imported, principally from the northeastern United States. Transportation still depended on lake navigation with all its attendant difficulties. Docks and supply bases, as before, were restricted to the lakehead and Sault Ste. Marie, and mines had to locate close to the lake. The government of the newly created province of Ontario committed itself to constructing colonization roads, however, so a reservation of 5 per cent of the acreage of lands was made for road allowances in patents issued after the silver discoveries of 1866.[3] The roads constructed in the lakehead region as a result of this action were generally short – half of them were shorter than six miles – but they were planned to form a network serving the mining interests by forming connections with the waterways, the Dawson Road constructed westward from Prince Arthur's Landing to Lake Shebandowa (1868-70) by the dominion government, and, eventually, with the line of the Canadian Pacific Railway.[4] These roads, like most of the others in the province, were almost impassable except during midsummer and midwinter.

A major local change that promoted the silver boom was the development of a regional supply centre capable of meeting the needs of the mining communities that were springing up between 1866 and 1875. There were two potential centres at the lakehead – Fort William and Prince Arthur's Landing. The latter, for decades a mere carrying place for the fur trade, was the quicker to respond

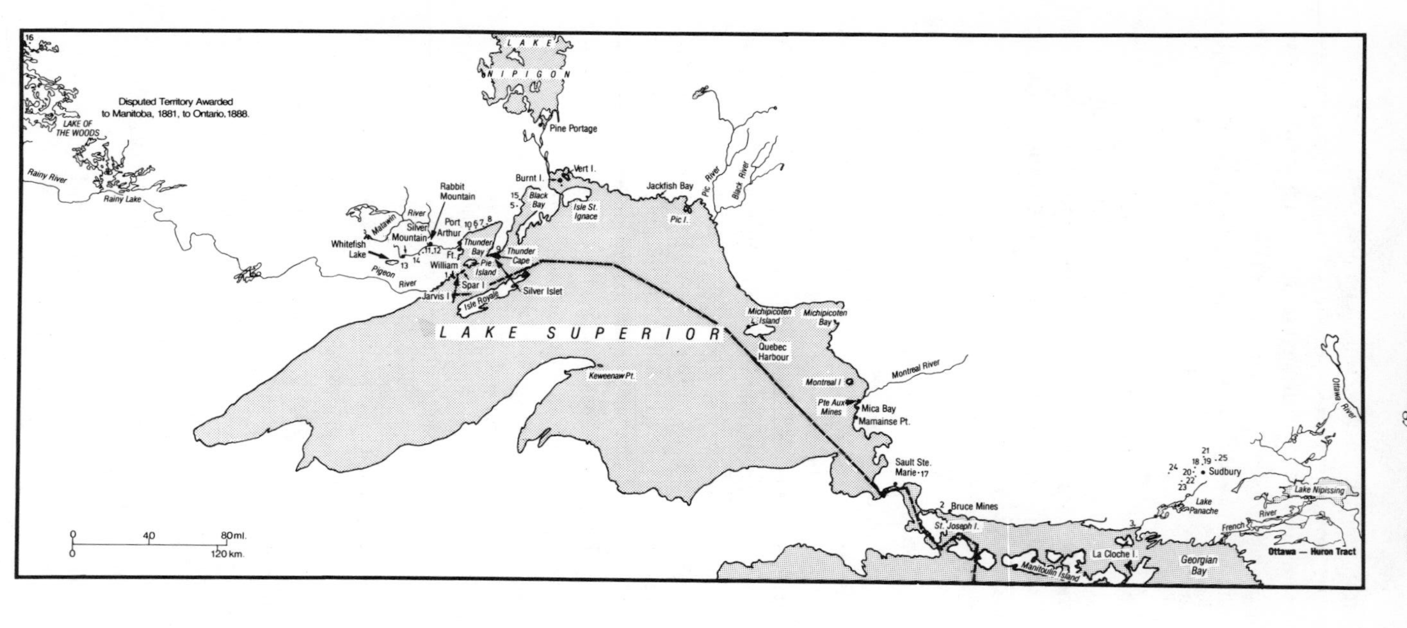

MAP 3 — Upper Great Lakes District

1. Prince's Mine
2. Bruce Mines (Copper Bay, Wellington, Taylor, Bruce)
3. Wallace
4. Quebec
5. Enterprise
6. Thunder Bay Mine
7. 3A Mine
8. Beck
9. Silver Islet
10. Shuniah (Duncan)
11. Beaver
12. Rabbit Mountain
13. Silver Mountain
14. Badger
15. Caribou
16. Sultana
17. Cascade & Victoria
18. Murray
19. Frood
20. Copper Cliff
21. Stobie
22. Evans
23. Creighton
24. Vermillion
25. Garson

MAP 3. The Upper Great Lakes Mining District.

in the late 1860s to indications that the region was rich in silver. By 1875, the Landing had become a boomtown. One of the most useful services it provided was weekly newspapers that generated mining publicity locally and farther afield. It was here, at the lakehead, that the province's first northern mining region evolved, nearly two decades before the celebrated Sudbury Basin opened to development.

EXPLORATION AND DISCOVERY PHASE, 1846-1850

The boom that occurred following the exciting copper discoveries in the early 1840s on the Keweenaw Peninsula of Michigan was the first United States mining boom and the findings were well publicized by the state geologist, Douglass Houghton.[5] A brief flurry of mineral exploration and speculation – but not much in the way of mining – occurred from 1845 to 1850 on the Canadian side, where the rich United States' copper-ore bodies were thought to extend. Prominent entrepreneurs and public figures from Montreal, Quebec, Toronto, and London, England, as well as Michigan mining men, displayed considerable interest in taking up mining locations.[6] The locations were originally obtained by individuals who then joined (or sold out to) Canadian and English mining companies formed to finance exploration and development. Since United States citizens were at first prohibited from taking up provincial mineral lands, or even from holding the majority of stock in British North American mining companies, they played no major role in the initial financing.[7] But companies frequently employed practical mining men with experience in the Michigan copper fields to locate mine sites and make assessments on engineering aspects and on the capital requirements – capital was scarce – needed to work those sites profitably.[8] Mine management, skilled labour, and technology, on the other hand, were imported almost exclusively from the long-established mining districts of Cornwall and Wales.

The only genuine boom that took place in these early copper years was in land sales by the new Province of Canada government, which was anxious to promote mining and secure revenue by selling mineral lands on a fee-simple basis.[9] Wholesale land speculation resulted, with Canadian companies apparently sitting on enormous mineral tracts until the day when conditions more favourable to profitable mining operations would boost resale value. Between 1845 and 1850, a total of sixty-eight locations in the Upper Great Lakes district were sold on installments.[10] These were generous tracts, two miles by five miles (6,400 acres); the short boundary was along the lakeshore.

Apart from a few seasons' work of preliminary testing and surface extraction on the islands and shoreline near Prince Arthur's Landing, on Michipicoten Island, and in the Mamainse region, including Mica Bay, the Lake Superior locations remained virtually idle until the late 1860s, when more favourable conditions arrived. The story of these pioneering ventures, however, offers some valuable insights into the serious difficulties which faced the infant Lake Superior mining companies in these years before the opening of the Sault Ste. Marie canal.

The first site worked in this early period belonged to Col. John Prince, of Sandwich (Windsor), who was a major shareholder in the British North American Mining Company, which had been formed in Montreal to operate Prince's silver mine.[11] After a year's preliminary exploration and testing, the company's mining agent, Alex Robertson, recommended that other working mines and various mining companies interested in Lake Superior be contacted to join in a co-operative shipping venture (a joint stock company) in the form of a steamer to visit places along the various north shore stations and Isle Royale.[12] He recommended also that the company petition the legislature for a canal to be built at Sault Ste. Marie. Letters were sent to Sir George Simpson of the Hudson's Bay Company, to the Montreal Mining Company, Quebec and Lake Superior Mining company, and the Upper Canada Mining Company, but only the last two companies expressed any interest. The idea of petitioning for a canal was soon dropped because of the poor state of government finances and the prospects of the United States government constructing one. The hoped-for capital and markets never materialized and after five years of preliminary work and highgrading that cost the shareholders about $5,000 annually, the company directors decided to abandon the operation.

The only other mining in the Lake Superior section was conducted on behalf of the Quebec and Lake Superior Mining Company, one-half of whom were Quebec timbermen.[13] The company controlled thirteen locations, all of them in the Lake Superior section, and conducted mining operations at Mica Bay and Michipicoten Island, which were closest to the Sault Ste. Marie supply post. William Palmer was mining captain for these operations, and his reminiscences, along with the observations of the Hudson's Bay Company trader at Sault Ste. Marie, reveal the serious constraints on mining in this region. In 1848, the company commenced its Mica Bay operation, where outcroppings of native copper had been worked since ancient times. Numerous permanent dwellings and mine buildings were erected, an overshot waterwheel and Cornish ore-dressing machinery were installed, and in the next year, a smelting works on the "Chilean principle" was to be added.[14] In a second attempt to overcome the problems of transportation in this district, a company backer, Allan Macdonell, actively promoted the construction of a railway across the portage at Sault Ste. Marie, but funding was not available for it.[15]

The Quebec company showed some initiative in pressing the government to negotiate a treaty with the Indians of the region, who were threatening to stop operations at Mica Bay if they did not receive compensation for their lands.[16]

Because the negotiations were slow and uncertain, the Indians temporarily seized the Mica Bay operation, holding it until the spring of 1850, by which time the Robinson Treaty had been signed. The company tried, without success, to sell the mine in England. Then the company abandoned Mica Bay in 1851, planning instead to develop its Michipicoten Island locations. But, when testing took place, the results were less than favourable, and the company abandoned this operation as well in 1853. The company had spent $100,000 for a disappointing total output of three or four tons of copper ore, all but exhausting the subscribed capital.[17]

Meanwhile, a thorough exploration of the north shore of Lakes Superior and Huron was conducted in 1846 on behalf of the Montreal Mining Company. Forest Sheppard, who led the exploration party, recommended the company acquire sixteen locations.[18] The company followed Sheppard's advice and in doing so became the major mining company for the district. But, unwilling to become pathfinders in the unpromising, largely unsuccessful Lake Superior section, Montreal Mining Company decided to confine development work to their most accessible Lake Huron location at Bruce Mines. The Upper Canada Mining company acquired a total of eight mining blocks, which, like those of the Montreal Mining Company, included sites in the Lake Huron section.[19] Within one year of exploration and testing each of its locations, it too decided to confine actual mining to one of its Lake Huron properties, the Wallace Mine, lying only sixty-hours-journey from home base in Hamilton.

BRUCE MINES, LAKE HURON, 1848-1876

The discovery of large, showy lodes of copper ore in quartz led to the opening of the Bruce Mine in 1848 and, as a consequence, to the first introduction of mechanically powered mining equipment in the province. The mine, located fifty miles below the Sault, was worked for two decades by the Montreal Mining company. In 1855 it was put on a sounder business basis by an English company that operated it, together with an adjacent mine, for a third decade. By the time production ceased in 1876, the workings extended for nearly two miles across the Bruce and Wellington Mines and the Huron Copper Bay Company locations, and a permanent village called Bruce Mines was established.

The Montreal Mining Company Phase, 1848-1855

While no original records of Bruce Mines and its experience with technological adoption and innovation have come to light, the story is detailed in the reports of the Geological Survey parties that visited the Lake Huron and Lake Superior locations every few years, in the annual reports of the Montreal Mining Company, and in the testimonies of successive mining captains and agents before the Ontario Royal Commission on Mineral Resources.[20] During its first season in 1847-48, numerous shallow shafts were sunk, levels driven, and some 1,475 tons of copper ore and undressed veinstuff were raised. Most mine work was by Cornishmen on the Cornish "tut" system, whereby the miners were paid according to the amounts of ore they raised.

In the first few seasons, the company was successful and Bruce Mines grew into a sizeable mining village with three frame and thirty log buildings and extensive shipping facilities for full-scale development, capable of accommodating several hundred workers and their families.[21] Seventy-seven miners, sixty-six labourers, an engineer, four boys, eleven blacksmiths, two clerks, plus carpenters and other artisans made up the original workforce. The mining superintendent, Captain Roberts, was a highly experienced mining man whom the company had brought out from England.

Despite these elaborate preparations, the operation was continually plagued with difficulties. The grade of copper in the ore from the mine was low, and the cost of freighting it to English markets high, which indicated the need to concentrate ores and smelt the concentrates before shipping.[22] Consequently, a powerful 45-horsepower Cornish engine on the Watt pattern, with a huge, twenty-eight-foot walking beam and two thirty-foot diameter balance wheels with massive rims of cast iron, plus ore-dressing machinery, including Cornish stamps, were installed in 1848.[23] A large Welsh smelting works, comprised of four reverberatory furnaces (three for roasting ores, one for smelting them) was erected shortly afterwards.[24]

The furnaces were fuelled with bituminous coal freighted by water from Cleveland, Ohio. The company hoped to recover the high costs of these works in several ways. First, in order to economize on fuel, the steam engine would be employed in as many tasks as possible: to drive the crushing rolls and buddles and to operate the sawmill and pumping engines. Once the shafts were driven deep enough, the engine could also serve to raise the ore.[25] In addition, the company hoped Bruce Mines would develop into a regional concentrating and refining centre for future mining operations and help recover the expense of the mill and smelter.[26]

Undertaking a full-scale operation at the Bruce was a formidable step, however, as the company's annual report for 1850 and the local Hudson's Bay Company trader's letters reveal. The remoteness of the region meant the company could take nothing for granted. Thus, in order to construct the furnaces, the company had to start its own brickworks. The engine house proved woefully inadequate to Canadian winter conditions, and it collapsed shortly after its erection, killing one man and disabling several others.[27] Second, with the fixed investment in place, the miners were in a good position to make demands of the company.[28] The miners hired in England on a monthly wage demanded a tribute contract so that they might benefit from the increased productivity. Then a cholera epidemic struck the workers, killing five and disabling the others. Toronto and Detroit proved the closest labour markets from which a temporary new force could be drawn. The new workforce from Detroit was German. They were supplemented by a number of former Western (Wisconsin) lead miners who were leaving the unsuccessful Mica Bay operations.

Company directors, responding to these bleak beginnings, attempted to place the Bruce on a paying basis in 1850 by dismissing the mining superintendent and hiring a new one under a five-year contract, also bringing in specialized English workers – including a furnaceman and three smelters. The company then ordered the erection of a new stone concentrating mill equipped with two sets of Cornish rolls. The ore was fed into the rolls by a large wooden wheel with wooden buckets along the inner rim. After crushing and sizing, there was a jigging stage. Here, crushed ore was passed by hand across coarse wire mesh over oblong pits to separate the ore from the quartz and other matter. One boy could attend three jigs. The dressed ore was then loaded onto carts and wheeled to a platform, thence into the smelter or straight onto vessels bound for Wales.[29] The mill was still intact when Archibald Blue visited it in 1892. Blue's guide was Sam Cullis, an old-time resident of Bruce Mines who had started as a jigger-boy and worked his way right up to the highest-skilled position, hammer man.[30]

Unfortunately for stockholders, the costs of recovery were too high.[31] First, where the Cornish technology proved useful for processing the high-grade copper rock in the Lake Superior copper region of the United States, it was unsuited to the task of treating the copper ores in this region. The ore was so intimately blended with the matrix rock and there was so little difference between the specific gravities of the two that insufficient copper could be separated from the rough ore to make the process pay. Moreover, the ore was found to yield only half the value that had been estimated when dressed. The second-class ore, which required repeated treatment and consequently high maintenance costs, proved too taxing on the pioneer resources in this remote district.

Secondly, the Welsh smelting works, like the milling system, was a disappointment.[32] The inefficient dressing methods left too much gangue mixed with the mineral, and most of the ore was of a refractory, or heat-resistant, nature, so that consumption of coal – a costly item in this region – was extremely high. As well, the Welsh process was highly labour-intensive; it required skilled operators who were not nearly as available in this region as in Cornwall. Cornishmen in the Lake Huron country were very independent, which often led to serious interruptions and even stoppages of the dressing operation when the mine manager attempted to pass along the operating losses to the workers. As a result of these difficulties, most of the officers resigned or were fired. In 1851, the newly elected directors replaced the mine manager with a Scotsman who had some experience in the Michigan copper field – Edward B. Borron. Borron, as mentioned earlier, led the mine through a brief period of rejuvenation.

When Andrew P. Salter and Count De Rottermond visited Bruce Mine on behalf of the Crown Lands department in 1855, they observed an operation that had changed considerably over the one witnessed in 1848.[33] Borron's first move had been to replace the former tut work system with a form of Cornish tributing, whereby the men were paid in accordance with the grade of ore they raised.[34] He had also constructed a new concentrating works for testing an "American" system of jigging recently developed in the Michigan copper mines and had the old stamps refurbished.[35] The American ore-washing apparatus must have been the Collom Washer, consisting of sieves travelling in continuous horizontal motion with added rocker motion. Borron also planned to replace the old Cornish smelter with a German-type blast furnace, using charcoal that would be produced locally. But by this time, 1852, copper prices were falling, and the company directors were losing interest in the smelting operation in view of the failure of mining to take off in this region and the high cost of smelting only their own refractory ores.

At this point, the operation entered into a phase of decline. Because Borron introduced the system of tributing, the terms of which were less favourable to miners in the case of lean ores than were those of tut work, most of the three hundred workers left the camp. Likely they headed for the Michigan copper fields or the goldfields of California.[36] With the mine force now reduced to a mere twenty men, the work had to be restricted to the richest, most accessible veins. Borron's strategy was to reduce company expenditures to "within safe and narrow limits," while at the same time keeping the ore-dressing equipment fully employed in cleaning the dumps and the tailings, or "skimpings," that had already accumulated on the site.[37] With the new ore-concentrating equipment, which was better suited to local conditions, Borron actually achieved a tiny profit for the Montreal Mining Company shareholders in the 1853 season.

The optimism fanned by this promising trend, the impetus provided by the United States undertaking of the Sault Ste. Marie canal, plus higher prices offered for copper, led the company directors to demand an escalation in production for 1853. This step called for an operational strategy of rejuvenation.[38] A larger workforce was required, which meant the popular tut work system was reinstated. The new men – mostly hired from Borron's native Scotland – were able to command high wages, comparable with those miners received in the more prosperous Michigan copper operations or construction workers were being paid for building the nearby Sault Ste. Marie canal.[39] Borron, however, blunted their demands for higher wages by guaranteeing them a four-year contract. He also increased the ore-dressing capacity of the site by purchasing a crushing wheel, which he had conveniently obtained secondhand from the unsuccessful Quebec and Lake Superior Mining Company operation, and also ten new jigging machines.[40] The workforce stood at 43 in 1853, double its 1852 level; and by 1854 it topped 140.

New problems ensued, as Borron's reports to the directors for 1854 and 1855 and his later testimony to the Ontario Mineral Resources Commission revealed. His attempts to increase production were nullified because the veins of copper were becoming noticeably thinner. The situation was aggravated by the breakdown of the company steamer in 1853 and the consequent failure of the new force of Scottish miners to arrive that season. Then in 1854, the company steamer, carrying a load of supplies for the winter, was lost during a storm. The cost far exceeded the loss of a ship and supplies; miners under contract still had to be paid, and replacement supplies obtained from Sault Ste. Marie were more costly. Besides, the ore-dressing apparatus was uneconomical. Thus, instead of producing increasing profits for 1854 and 1855, the expanded operation showed substantial losses. The company decided to abandon the shortlived policy of rejuvenation that entailed continued reliance on new technological innovations and employment of a large mining staff; instead, it intended to reduce the operating losses by reducing the scale of work.

In the attempt at least to break even and perhaps even make a small profit, the company chose to get out of the mining business and instead to support what would become known in the entire Upper Great Lakes district as "scramming" operations.[41] Under this system the miners worked on their own account, paying for their own supplies and the wages of the ore-dressers and returning a percentage of the value of dressed ore to the company. The company, as supplier of mining materials and the accommodations and as shippers, occupied a monopoly position; it stood to earn a profit from the mining activities even if mining proved unprofitable. Because it was in the company's best interests to protect its investment in the site, it agreed to maintain the underground works and the ore-dressing equipment and to keep the engine, pumps, and ore-dressing machinery repaired and running. It also lowered the rents on the company houses if the miners agreed to perform the necessary repairs to them.[42] Though the existing contracts still had three years to run, they were cancelled with the support of the miners in favour of the new contract and leasing arrangements.

The West Canada Company Phase, 1855-1876

The entire western portion of the property had been leased to a Cornish miner on a royalty (5 per cent of the value of the dressed ore) basis in 1853.[43] Sampson Vivian quickly organized an English company of operators known as the West Canada Company and put it under the management of the prominent English mining engineering firm John Taylor and Sons. Accidental discoveries of veins resembling, but even richer than, the original discoveries resulted in the full-scale development of a new mine, called the Wellington, in 1859. Because the rich veins appeared to extend to the adjacent Huron Copper Bay Company location, that too was leased.

When the provincial land surveyor William Gibbard visited the Upper Great Lakes mining district in 1860 (after twelve years of mining activity there), he found little to report.[44] The Lake Superior section, which contained the bulk of the locations, was entirely inactive. Indeed, the ownerships of most of the original locations were about to be forfeited.[45] In the 1860-61 season, a Mr. Fletcher tested one of the Quebec and Lake Superior Mining Company properties for prospective buyers in New York City. He erected a steam plant and a small stamp mill, only to run into financial difficulties that, together with the fall in the world price of copper, resulted in the decision to abandon the testing operation. The bulk of the company shares eventually fell into the hands of a few parties in Quebec. When their lumber business failed, the mining venture also foundered.[46]

On the north shore of Lake Huron, sixteen mining properties, several comprising two or more locations, had been granted over the years, but mining operations had actually been conducted only on two of these – the Bruce Mines and Huron Copper Bay. Tiny trial openings had been made on nine others (for example, in 1848 at Wallace Mine), but by 1862 they were idle or abandoned; no work whatever was done on five of the number.[47] Gibbard also reported that miners were leaving the region. Some, he thought, were returning to England, while others were leaving to join in the United States Civil War. While this may well have been the case, it was also true that the high wages offered by the Michigan mines at this time lured several hundred miners from the Canadian side.[48]

At Bruce Mine proper a few dozen individuals had been conducting mining on their own account since 1855. Gibbard observed that under this leasing system a miner necessarily removed only the best paying ore, "merely severing a passage in places for his body, and following the most accessible veins, rather than the main lode."[49] He recognized, as his Geological Survey colleagues had done earlier, that this system was ultimately wasteful. So much dead rock was accumulated and so little development work was undertaken that the owners' chances of developing the property into a paying proposition or even of selling it were seriously compromised. The big advantage to the owners, however, was that very little supervision or expenditure was required and that the equipment was kept in working order.

Unlike the Bruce, the Wellington Mine was in the heyday of its prosperity at the time of Gibbard's inspection in 1860. The Wellington employed 330 hands,

125 of whom were underground miners (mostly Cornishmen) and supported an even larger population, constituting in all over half the entire mining population of Ontario. The miners were paid at the going rate, between $32 and $35 per month, by contract, while the surface labourers were paid a typical daily wage of $1.00. Shafts reached levels deep enough to require the use of steam pumping and hoisting engines. Steam was also used to drive the crushing apparatus, which consisted of Cornish rolls rather than the modified Cornish stamps once popular in the Lake Superior copper region.[50] The concentrates were shipped to Baltimore and England for smelting. Drilling underground, as was standard for the period, was by hand.

It was William Plummer, an experienced English mining captain who had previously managed a number of mining operations in England for Taylor and Sons, who reintroduced a strictly Cornish system of mining to this area.[51] Plummer testified before the Royal Commission on Mineral Resources that he had brought with him the most modern mining and concentrating apparatus that English experts considered appropriate for successful, full-scale development of a copper deposit. As always, the freight and the start-up costs for such an undertaking were high in this region, requiring a large outlay of capital (about £200,000) and delaying the earning of profits. As at the Bruce the quality of the copper ore found at these depths was low (approximately 5 per cent). Too much copper was lost under the contemporary English system of concentrating, so the relative richness of these newly discovered deposits was insufficient alone to guarantee successful operations. In fact, no dividends were realized by this company for almost a decade. Unlike the Montreal Mining Company shareholders, however, those of the West Canada Company were willing to wait.

The total reliance on Cornish technology and skilled labour in this remote region once again gave a high degree of independence to the miners. When management attempted to impose new hours of work during the winter of 1864-65, a riot brought the cancellation of the offensive regulation. The company, however, regained some control by dismissing thirty of the "most troublesome" men.

At this time in the mid-1860s, the operation was finally showing a profit, and the company of operators decided to attempt a major rejuvenation. Vivian's lease was about to expire. In 1865 the world price of copper fell to crisis levels, and the proprietor, the Montreal Mining Company, was in the process of reassessing its holdings and had largely abandoned any hope of developing the area itself, so it sold the property to the operators.

The new owners erected two mills of successively larger scale and employing Cornish rolls in the next few years. The rolls, engines, and boiler were all imported directly from England.[52] John Taylor visited the property in 1868 to examine the operation and plan for its future direction. Like Borron earlier, Taylor noticed the indications of a falling-off in size and richness of veins, even at shallow depths.[53] His principal recommendation – the immediate installation of a reduction furnace – was a return to the earlier technological solution of Borron's day. This time, however, a new, experimental technology that was thought more in keeping with the local conditions of high fuel and transportation costs than the Welsh method was proposed. In the new smelter milled ore was roasted in brick reverberatory furnaces, ten in all, then the roasted ore was crushed, combined with salt, and re-roasted. After this, the calcined ore was washed and leached in hot, dilute sulphuric acid; the resulting leach liquor was drained into wooden vats containing scrap iron to produce a 70 per cent copper cement for smelting.[54] On-site smelting was now necessary, since the high United States tariff had effectively cut off imports of foreign copper ores for smelting there.

The attractive features of the new furnaces were their relatively low fuel requirements and their good performance in treating low-grade ores. The new technology, which had been developed in Scotland, appeared better suited to the needs of copper-mining in this region than the older Welsh system of smelting attempted at the Bruce. Unfortunately, this new system required large amounts of scrap iron and salt – commodities which, like coal, were scarce and costly in this region. The difficulties with supplies made it simply unprofitable to treat the lean ores. The reduction works, for which the company had spent $100,000 and which took from 1869 to 1871 to complete, operated for only two years, and at the same time a third concentrating mill, of much larger capacity, was erected in the centre of the village.[55] With the economic crash and falling copper prices in 1873, once again the smelting operation, or "chemical works" – intended to prolong the life of mining – was the most expensive unit to operate, and was the first to be shut down.

Despite the commendable amount of technological experimentation and the apparently successful production at this location, the difficulties for mining persisted. With the price of copper continuing downward, with rapidly depleting mineral resources and no new discoveries, and with the loss of its copper markets caused by the dramatic ascendance of Michigan copper (lead by the Calumet and Hecla Mines), mining operations ceased altogether at the Bruce Mines in 1876. During its period under the Montreal Mining Company copper ore valued at $3.3 million was taken out, most of it being shipped to Swansea. Following that, the English company extracted between $6 million and $7 million worth of copper.[56] Still, this output was insignificant compared with the output of the Michigan copper region, which by 1860 was producing copper ore worth $14 million annually. Meanwhile, the focus of attention in the Upper Great Lakes had shifted from Lake Huron to the Lake Superior section.

THE LAKE SUPERIOR SILVER PHASE, 1865-1884

Gold discoveries and a general surge of activity by the mining industry elsewhere in the province led to a succession of acts being passed in the 1860s that attempted to discourage mining speculation and encourage active development. For instance, the crown reservation on gold and silver was expressly abandoned, the size and cost of acquiring new mining locations in the Upper Great Lakes district was reduced, and a tax based on acreage was imposed on mineral lands that made it more costly for companies like Montreal Mining to retain large holdings of mineral locations idle.[57] These new regulations as well as the improving economic conditions – especially the lowering of transportation costs – led to renewed interest in the Lake Superior section.

Renewed explorations resulted in promising discoveries of native silver, silverglance, and galena.[58] The Back Bay lode (Enterprise Mine) was discovered in 1865, followed over the next few years by even more important discoveries – the Thunder Bay, Shuniah, and Wallbridge mines. Moreover, silver was found on Victoria, Pie, and McKellars Islands and, further east, in the vicinity of Pic and Black Rivers. Those responsible for the discoveries were local prospectors and Hudson's Bay Company personnel, usually acting on tips provided by Indians who lived and worked in the area. All but two of the Montreal Mining Company's sixteen Lake Superior locations were in the Thunder Bay area, where the best finds were being made. Consequently, in 1868 it dispatched geologist Thomas Macfarlane and a large exploration party to evaluate its mineral lands for future sale. In that first year Macfarlane uncovered silver on Jarvis Island and the large, rich lode on Silver Islet.[59]

Not all the difficulties facing mining in the Upper Great Lakes district were overcome by the new discoveries of silver. Indeed, silver-mining operations got off to a rather slow start, because silver ore is a much more highly complex mineral than copper. Exploration proved difficult because the mineral occurred in rich surface showings that quickly ran out and in veins that ran quite deep, rather than in convenient, easily worked deposits. The silver veins were difficult to predict because they occupied irregular cracks in rocks.[60] All in all, silver was hard to detect, mine, or dress without a great deal of technical equipment and, therefore, capital. Fortunately, the appropriate technology and expertise were being developed in the United States during the 1860s and 1870s on the south shore of the district and in the American West.[61] Moreover, United States markets and capital were favourable to exploitation of the Canadian silver field during the brief economic boom period that followed the Civil War.[62]

The phenomenon of American capital developing mineral lands that a Canadian company would not exploit became notorious in the Thunder Bay area, and it was subsequently regarded by local newspaper editors as proof that

United States' capital and expertise were requisite to Canadian mining development. Canadian, and some English, capital had been poured into the Lake Superior silver boom, but it was principally confined to the speculative, rather than the development, phase. The Canadian-owned Wallbridge Mine had been opened in the late 1860s; a shaft was sunk – but only forty feet. Its owners were local men from Fort William and Sault Ste. Marie, and when the shaft filled with water, they abandoned the site for want of capital.[63] Another Canadian-owned mine, Pie Island, underwent a similar fate. Superintended by its discoverer, John McKellar, the mine limped along for years, but in 1882 it was suddenly described as "a hive of industry."[64] Then came a report that men were quitting over pay problems, and in November supplies failed to reach the mine. Operations there were abandoned too. Peter McKellar's Thunder Bay Mine was soon to furnish yet another example of a Canadian failure. The McKellar brothers, John McIntyre, and Governor Hopkins of the Hudson's Bay Company had formed a Montreal-based company in the late 1860s.[65] After securing 1,700 acres of mineral lands and spending £60,000, operations ground to a halt from what the Thunder Bay *Sentinel* termed "a lack of vigor." Then, in 1874 a New York syndicate took over the operation, brought in the necessary equipment and workforce, drained the shafts, built roads, and soon had successful production underway.

Superintending the new Thunder Bay Mine operation was an American, Alex Stewart, who was also involved with rejuvenating the old Duncan Mine. The Duncan had been worked only sporadically and without much success as a small-scale operation known as the Shuniah Mine until 1873, when new United States owners undertook serious testing with a diamond drill in 1877.[66] In all, the new owners spent about $200,000 on this site, but to no avail. However, for his work in reopening idle mines and his promotion of a concentrating mill at Prince Arthur's Landing, Stewart himself became something of a hero in the Thunder Bay region.[67]

Perhaps the most dramatic example of American capital developing mineral lands that a Canadian company would not exploit was Silver Islet itself, which was known to be extraordinarily rich from the beginning. Thomas Macfarlane, its discoverer, tells the story of his test-drilling on the Islet and his attempts to interest the directors of the Montreal Mining Company in the silver deposit.[68] Not willing to develop the property itself, the company spent two years in vain attempts to interest English capitalists. Ultimately, a New York-based syndicate, the Ontario Mineral Lands Company, headed by veteran Michigan mining men, bought out the company in 1870, paying $225,000 for the Lake Superior locations (roughly 107,000 acres), including Silver Islet. Immediately following their purchase of the Montreal Mining lands, the new owners dispatched a company manager, William Bell Frue, with thirty men, horses, and a boatload of machinery and supplies. Upon their arrival, they fell upon a surprised Macfarlane,

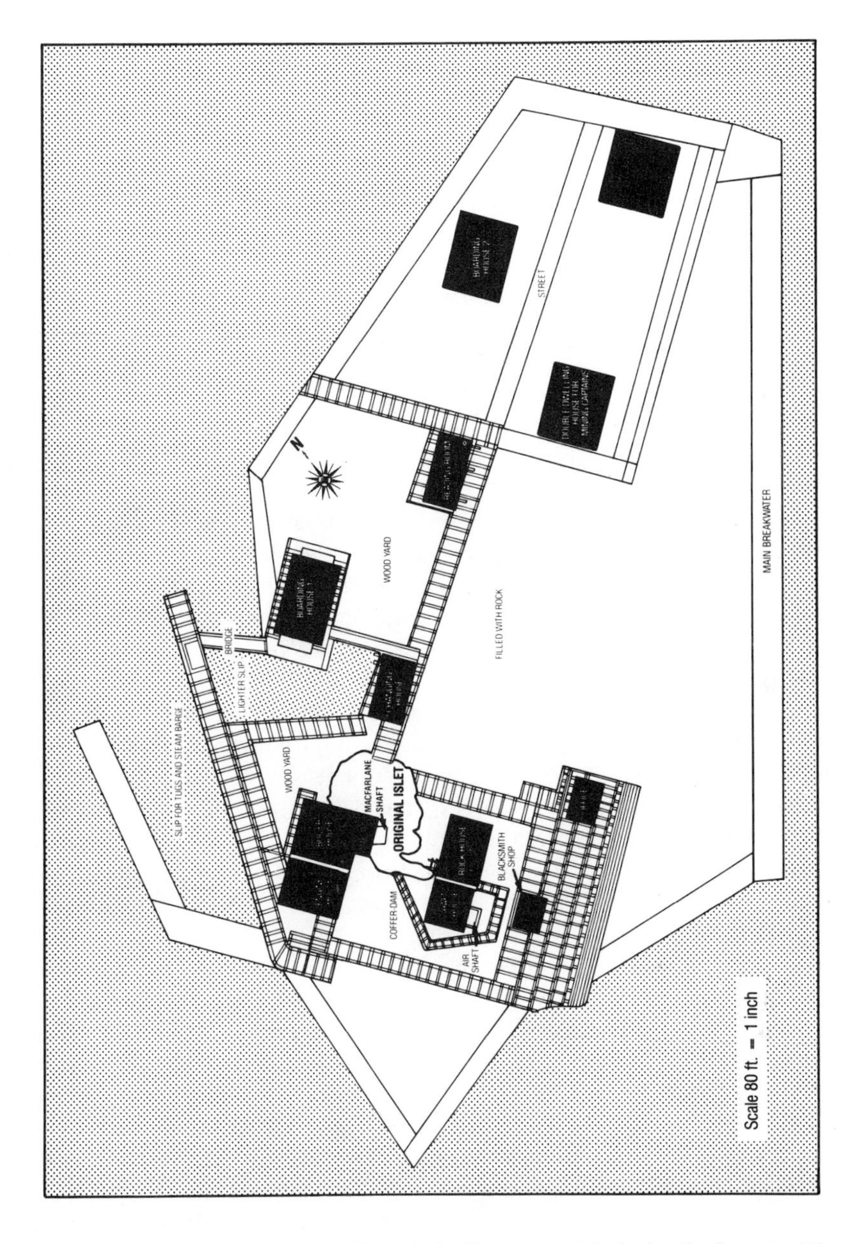

Figure 4. Plan of Silver Islet in 1879. *Silver Islet itself was a mere eighty by sixty-five foot outcropping located about a mile off Cape Thunder and rising only eight feet out of Lake Superior. The Silver Islet company spent $250,000 in the first year erecting and maintaining a huge coffer dam and elaborate system of breakwaters around the original islet and developing this site and the townsite of Silver Islet Landing on the mainland. Accommodation for the miners was erected right on the Islet in order to keep mining going on during the periods of inclement weather.*

still chipping away at the rich outcroppings, unaware of its sale. Macfarlane had advised the Montreal Mining Company that to begin development of the Islet, $50,000 would be required; the Silver Islet company spent a full $250,000 in the first year, erecting and maintaining protective works – a huge coffer dam and elaborate system of breakwaters – around the tiny islet and developing the townsite of Silver Islet Landing on the mainland.[69]

While United States capital and management may have played the most significant role in the Thunder Bay silver region, American labour did not. Cornishmen and Norwegians were the nationalities most frequently represented in the workforce. Few miners had been employed in the earlier copper-mining era, and with no large permanent settlement, most of the labour required had to be brought into the area from outside. Fortunately, the silver boom coincided with a massive emigration of miners from Cornwall as mines closed down there.

There is almost no record of the working conditions in the mines of Lake superior, but in all probability they were similar to those in hardrock mining districts throughout the Western world. According to the few surviving eyewitness reports, the men worked ten-hour continuous shifts if at the surface, eight-hour shifts underground.[70] There was no work on Sundays. The underground force laboured in damp, cold shafts illuminated only by candles stuck onto their caps or jammed into the rock face (the miners paid for the candles and also for the blasting caps, fuses, and powder). Communication between the levels and the surface was accomplished by a primitive bell-signal system. The mining labourers working at the surface included rock wheelers, bucket handlers, timber boys, and tool-sharpeners. Boys worked in the rock house, breaking up the ore, carting it, and packing the high-grade material. Then there were unskilled labourers, including a few Indians and "boys," and skilled craftsmen, all of whom worked at various tasks – chopping and hauling wood; sawing lumber and making or repairing barrels, ladders, sleighs, and wheelbarrows; building roads, buildings, wharves, breakwaters, and other structures; operating and repairing engines, pumps, and machinery; and running the concentrating mill. A host of labourers were kept busy just scouring for rock and timber for the protective works. And dockmen, watchmen, and sailors were needed as well.

Accommodation for workers was provided by the companies, and normally it consisted of boarding houses for single men and tiny frame houses for families. Approximately half of a miner's monthly pay went for room and board. At Silver Islet, three boarding houses – one for the Cornishmen, one for the Norwegians, and the last for "others" – and a double-dwelling house for the mining captains were built right on the Islet itself, apparently in order to keep mining going on during the winter months when Lake Superior was frozen solid and the men experienced difficulty in travelling from the mainland.[71]

Camp discipline was strictly enforced at all the mines. Camps had their own lock-ups and restricted the consumption of alcohol. Camp watchmen kept strict vigil, in part to maintain law and order, but mainly to keep watch for fires. The companies provided far more than mere shelter and discipline for their workers; many mining camps contained doctors and clergymen too. Silver Islet possessed a hospital, school, and two churches (although there were three clergymen). There were, in addition, a reading room, saloon, and search house on the Islet, and a saloon, jail, company store, customs office and bank, officers' housing, and forty workers' houses at the Landing, which by 1877 stretched a mile along the lakeshore. Last, the Lake Superior mining companies sponsored a baseball league for their workers, an activity in which Silver Islet, as in all other endeavours, was leader.

Silver Islet Mine, 1870-1884

Silver Islet was the crucial underpinning for metal mining in the Upper Great Lakes district, as the earliest and ultimately the only bonanza. Thus, it deserves close examination as a "leader" operation. Silver Islet was an anomaly; it was the one mine in the province to achieve a fame in its time that ranked with the best silver mines of the American West.[72] And it was here that the major technological advances and mining strategies of the day received their regional trials.

The Islet, a mere eighty by sixty-five foot outcropping located about a mile offshore and rising only eight feet out of the Lake, contained an extraordinarily rich ore body. The Silver Islet mineral occurrence comprised a series of rich bonanzas in veins and deposits, separated by ores of lesser value and much worthless rock. Some rich ores went for as high as $10,000 a ton, while other ore hardly paid the expense of extracting it. Because of its position and small surface dimensions, the Islet shaft had to be driven deep, so the operation led the region in deep-level mining, ultimately to a depth of 1,260 feet, with fifteen levels of tunnels intersecting the main shaft, from which cross-cuts driven across the course of the veins extended far under the storm-tossed lake.

The richest ores were uncovered near the surface in the first few years, following which efficient concentration of low-grade ores sustained operations for a lengthy period. No expense was spared, especially during the first ten years of its operation. The economic recession that occurred in the mid-1870s and the tailing off of the rich veins brought about a cessation of work at the Islet for one season. Prospects were so good, however, that the company reorganized and started up operations once again, this time as the Silver Islet Consolidated Mining and Lands Company. The rejuvenation was possible because of a $400,000 mortgage on the property. The investment paid off a few years later, when a rich bonanza was struck in new drifts at the four hundred-foot level.

No other mining operation in the province during this period proved promising enough to receive the amount of time and money invested in plant and equipment, especially new technology, than did Silver Islet. In its first year of operation, 1872, the Silver Islet company brought into use a battery of the new Burleigh compressed-air drills.[73] It was one of the drill's first applications to mining, and new drills made possible more thorough and speedy exploitation of the deep-running veins, reducing the previous wastage of paying mineral ore.[74] These drills affected the workforce considerably because the drills also freed mine managers from their almost total dependence upon the original workforce of traditionally skilled Cornish miners, who had been recruited in Britain by the Islet's Cornish mine captains, the brothers Richard and John Trewtheway.[75] The mineowners, who were perhaps alarmed by a major labour strike in the Michigan mines, recruited less-skilled, less-independent, cheaper mine labourers such as were readily available at other North American mining camps in Nova Scotia, Lake Champlain, and Galena, Illinois.[76] "If miners are plentiful, make up the number from mixed nationalities," read Frue's instruction to William Savard, who was the company agent.[77] By the end of 1873, a heterogeneous workforce of three hundred men was employed at Silver Islet.[78]

William Frue's letterbook reveals how managers planned to impose stricter control over this new workforce and to break down the independence the Cornish miners had displayed.[79] However, using such a heterogeneous workforce increased accidents, and clashes among workmen were frequent. The Norwegians were particularly troublesome: some abandoned their jobs; others went on strike. In turn, the management at Silver Islet was led to impose stricter work discipline and to seek additional productivity-raising devices. The ultimate threat to the workers, as before, was dismissal, though now men had only to go to Prince Arthur's Landing to learn of work prospects at other mines. By the same token, the growing pool of mine labour at Prince Arthur's Landing obviously served the interests of mine operators: it was easier to dismiss workers if they could readily be replaced.

The use of compressed air in mining spread quickly from Silver Islet throughout the district. A drill similar to the Burleigh was first tested and used in 1880 at the neighbouring Duncan Mine, the most successful mine next to Silver Islet. Built by the Johnson Rock Drill Manufacturing Company, this drill was experimental, and it was appropriately named the "Duncan" (although the miners themselves referred to it as the "Mrs. Johnson" because of its diminutive shape).[80]

A second sort of technology was introduced to exploit the less accessible veins and deposits lying at greater depths. As shafts were driven deeper and deeper, greater amounts of water had to be pumped correspondingly farther distances to the surface. Canadian manufacturers were able to make a contribution in this one area. Although steampowered pumping engines of United States manufacture were advertised in the local newspaper, the mines in the region

Plate 8. The Silver Islet Shaft House (background) and Rock House, c. 1870s. *Ore from the mine was rough-sorted and broked up in the rock house, then the low-grade ore was transported to the company's concentrating mill, located on the mainland.*

tended to use boilers and engines made in Canada, especially ones from Toronto's large Soho Foundry and one in Peterborough.[81] A modest, single forty-horsepower steam pumping engine was installed in the first year, but others of successively greater capacities were added every few years. In all, four generations of increasingly powerful pumping engines were installed at the Islet, interesting testimony to the success of the mining operation and the main reason for the mine's ability to advance to deeper-seated ore. So much installation work needed to be done at the Islet that one employee, Adam Neilson, made this work his speciality and started an extra business laying engine foundations and erecting mill buildings throughout the Lake Superior section.[82]

Continuous shifts of engineers tended the pumping unit, and a machine shop was maintained where repairs could be made and new parts manufactured. Engine breakdowns were usually the result of poor maintenance or overwork and had to be prevented, for without a fast, reliable large-capacity pumping unit, the shafts would quickly fill with water that would require months to empty. In addition to the engines themselves, the rods and connecting chains frequently broke. In general, the engines (and rock drills) required constant maintenance and frequent repair.

Of greater lasting importance for mining in the district was the company's progress in smelting and refining the metal ores. In fact, the company was the first one on the Canadian side to overcome the early difficulties in treating the district's ores. Like a few other companies before them, the Silver Islet company quickly invested in an enormous smelting and refining works. But where the various attempts at Bruce and Wellington mines had failed, this operation succeeded for two important reasons. Firstly, it successfully combined a variety of American, Welsh, and German smelting practices. Secondly, its works were conveniently located in Wyandotte, Michigan, where they could handle a large volume of mixed Western and Lake Superior ores, lowering the fuel consumption per unit.[83] The Western ores contained recoverable amounts of lead, which were smelted along with the silver in low-blast furnaces (Krummöfen). In addition, the nickel content of both types of ores was recovered and refined in reverberatory furnaces, and even refuse too poor for smelters was treated to make it commercially valuable.

The company also proved that lean ores could be worked profitably during periods when mines no longer were producing high-grade, or "packing," ore. This was a stage faced by all mining operations in this era. After the rich ore of the first deposit ran out at Silver Islet, company manager Frue turned his attention to the low-grade mill rock which had accumulated in dumps on the property.[84] The decision to upgrade formerly unmarketable ores to a richer concentrate for shipping to the smelter led to the combination of a number of existing innovations in milling equipment between 1872 and 1874 to meet local conditions. An amalgamation process used in Idaho on silver ores was tested, for example, but it was abandoned in favour of a concentration process, perhaps because of Frue's long experience in the dressing of Lake Superior ores.[85] The result, as described earlier, was the patented invention of an improved mechanical ore separator, which was called the "Frue Vanner," after its inventor.[86]

The Frue Vanner was employed at the Silver Islet operation in conjunction with the otherwise typical Lake Superior concentrating mill. Silver Islet employed twenty-four Vanners from 1875 onward, housing them in an impressive $90,000, five-story mill along with four steam boilers, a 250 horsepower engine, three friction hoists for raising ore into the mill, two Blake crushers, and ten batteries of Fraser and Chalmers steam stamps.[87] Low-grade ores were broken up in the crushers and ground in the stamps, then mixed with water and dressed so that the silver particles were separated from the rest by sifting. Dressing was accomplished by means of the Frue Vanners. Each of the vanners could process in excess of six tons of ore every twenty-four hours.

Its capacity was not the Frue Vanner's main contribution, however; that lay in its high recovery of fine particles of silver from slimes. The ingenious new sifting device was capable of concentrating formerly worthless low-grade ores into a concentrate with a 90 per cent silver content at the low cost of $150 per ton of original ore processed. As a result, the Vanner greatly extended the life of the Silver Islet Mine and, farther inland, that of three of the Thunder Bay area's most promising operations – the Duncan, Rabbit, and Silver Mountain mines.[88] Furthermore, it gave tremendous encouragement to continued explorations and the development of gold-mining locations inland in the disputed territory.

In addition to its local impact in the Lake Superior section, the Frue Vanner was a considerable contribution to the world of metallic-mining technology in treating various lower-grade slimes. It won serious recognition, for example, in England, where it received favourable trials as early as 1875 and was considered as a possible solution to the problems of the ailing Cornish tin industry.[89] A description of the Vanner was carried in the leading mining journals, and the Fraser and Chalmers Company – which bought the patent and hired Frue – introduced Frue Vanners into the new mining camps of North America, including lead, copper, and silver fields in the American West, the Alaskan goldfields, and other mining regions around the world.[90] The English manufacturer of the Frue Vanner, T.B. Jordon Sons and Commons, distributed it internationally as part of their Continuous Ore-Dressing Plant.[91] And a decade after its invention, mining engineers at leading North American schools, such as the Massachusetts Institute of Technology and the California College of Mining, were being trained in its use.[92] Frue's invention was the ultimate development of the (mechanical) type of concentration process and a fine example of a local invention.

Despite its popularity, however, the Frue Vanner seems not to have been adopted in the Michigan copper mills. There, the Evans revolving slime table, which had been developed about the same time to deal with the problem of losses in dressing the Michigan copper ore, was already being used in practically all the Michigan copper mills. That is not to say that the Michigan operators did not experiment with the Vanner, for at least one trial occurred, but the Vanner simply proved not to be more economical than the Evans Table for treating the copper ores of Michigan.[93]

Still another major technological innovation introduced by the Silver Islet company was the diamond-tipped exploration drill, which permitted operators to test the nature and character of prospects and mines fairly scientifically. Like the compressed-air drills, the diamond-tipped drills were new on the market. Their introduction and immediate success at Silver Islet in 1873 was surprisingly swift.

The portable steampowered diamond-drilling rig introduced at the Silver Islet operation was used specifically for prospecting purposes.[94] The diamond drill proved to be of substantial benefit to the operators of Silver Islet and surrounding operations: first, it demonstrated the relative worthlessness of the other mineral locations held by the company, enabling them to concentrate their efforts at the Islet. Secondly, it revealed previously undisclosed bodies of ore underground that revived flagging operations during a time of economic depression in 1877.

The Duncan Mine employed a Bullock diamond drill shortly afterwards; it was supplied – probably for promotional purposes – by the Chicago office of the Fraser and Chalmers Company. Lavish advertisements for the Bullock Drill appeared regularly in the Thunder Bay *Sentinel,* while, in the tradition of the day, M.C. Bullock himself was reported to have visited the region from New York to inspect his drill in use there.[95] It proved unsatisfactory, however. Worked for a full eight hours, it consumed a complete bit and produced a mere six-inch core of rock.

No details have come to light concerning any similar difficulties in employing and maintaining the diamond drill at Silver Islet. However, clearly great uncertainty surrounded deciding the location, depth, and number of test holes required to produce accurate results for specified kinds of mineral deposits and structures and for the economic aspects of expenditures. A decade later, the operators of the Beaver Mine reported their diamond drill satisfactory for determining the configuration of a lode, but they would not accept the diamond drill results as definitely condemning a vein, only as locating a master vein. In other words, fifteen years after it came into widespread use, there still were no clear rules for the practical and economical use of the diamond drill.[96]

Production at Silver Islet fell off after 1880; no new rich deposits were being uncovered at the Islet mine or on any of the company's other properties, and the men with the greatest interest in the operation — A.H. Sibley and William B. Frue — were dead.[97] The company's strategy during the period of decline had been to sell off some of its assets — the tugboat, laboratory equipment, and even its ore bags — and to confine underground work to the most accessible (upper) part of the mine. With the worldwide economic downturn in 1883, the company found it impossible to assemble the necessary capital to keep the operation stable, so when the winter's supply (1,000 tons) of coal was lost in a shipwreck in 1884, the pumping engines were shut down, the main shaft was allowed to fill with water, and the operation was abandoned forever.[98] Over its roughly fifteen years' operation, Silver Islet yielded three million ounces of fine silver with a market value of $3.5 million. The property would be re-tested several times in future years, but never with encouraging results.

Plate 9. Silver Islet Abandoned. *Production at Silver Islet fell off after 1880. When the winter's supply of coal to fuel the pumps was lost in a shipwreck in 1884, the operation was abandoned forever. Over its fourteen years' operation it yielded three million ounces of fine silver. The property was re-tested many times over the years, but never with encouraging results.*

BEGINNINGS OF THE NEW NORTHERN MINING ERA

By the time the Silver Islet operation entered the decline and abandonment stage, the entire situation for mining in the Upper Great Lakes district had changed. The construction of the Canadian Pacific Railway section across Northern Ontario, together with subsequent lumbering operations, ushered in a new mining era for the region.[99] The building of the railway line coincided with the economic resurgence of 1880-83; its impact on mining was swift. Where the line of construction followed the north shoreline across Lake Superior, several of the old mining properties were retested and some of them even were opened to small-scale development. Then there were a dozen new mines discovered slightly inland from Sault Ste. Marie, at Echo Lake, where iron ore and lead-silver ore were mined, and west of Prince Arthur's Landing, at Rabbit Mountain and Silver Mountain, where important discoveries of silver were made. Finally, there were discoveries of gold in the Lake of the Woods area, the disputed former Hudson's Bay territory lying between Ontario and Manitoba, and the discovery of the rich nickel-copper deposits in the Sudbury Basin.

Of the old shoreline properties only two showed particular promise. The first one, Mamainse Point, was worked from 1882 to 1885 by its new owners, the Lake Superior Native Copper Company.[100] The company invested lavishly in the initial development, installing winding engines and compressed-air drills underground and erecting an old-style Lake Superior concentration mill, with Ball stamps and Collom jigs, a sawmill, and a machine shop. A large workforce of 260 men was employed, of whom 75 alone worked as choppers. Miners worked by contract, and labourers were paid relatively low daily wages of between $1.25 and $1.50. Superintending the operation was J.H. Trewtheway, the Cornishman who had begun his career in this region as a mining captain at the Silver Islet Mine. By 1885 it was clear to the stockholders that the operation was grossly overextended. The company went into liquidation shortly thereafter.

Attempts to reopen an old mining property on Michipicoten Island met with a similar fate.[101] In addition to the usual problems of isolation and uncertain supplies of ore, the English owners experienced difficulties raising the necessary capital at home and managing the operation from such a distance. As with the Mamainse Point operation, a great deal of development work occurred, and a mill, this time equipped with Cornish rolls, "plunger jigs," and buddles, was constructed. But the operation never actually advanced to the point where the owners felt they could afford to dress or ship the stockpiles of ore on the site. A new English owner tested the property thoroughly with a diamond drill in 1887, but when he died, there were problems selling off the property, given the isolation of the location and the poor economic climate.

There were slightly more encouraging developments at Echo Lake, where deposits of argentiferous galena (silver and lead) were worked at the Victoria and Cascade mines.[102] These two mines were located close to each other and only a few miles from the planned Sudbury-Sault Ste. Marie branch of the Canadian Pacific Railway line (completed in 1887). A considerable amount of development work occurred, and a small steampowered stamp mill was erected at each. Steam was also employed in hoisting. Labourers here were paid the same low rates as at Mamainse Point. These low wages were likely the result of the proximity of these mines to the supply of cheap labour available at the Garden River Indian reserve. At the Victoria Mine alone, thirty to forty Indians worked chopping wood in the winter months. Iron-ore mining occurred as well in this region, and, in fact, iron deposits were being discovered throughout the entire Upper Great Lakes district, but usually they were not in a form or quantities that paid anyone to work them.[103]

Of much more significance than these mining developments on the north shore of Lake Superior were the exciting discoveries of silver back of Prince Arthur's Landing in the vicinity of the Canadian Pacific Railway section stretching from the Landing to Winnipeg. The discovery of silver at Silver Mountain in 1882 caused the same type of rush to the area as had the Thunder Bay discoveries fifteen years earlier.[104] A number of mines were started up with the assistance of Indian labourers from the Sturgeon Lake reserve; the most significant were the Beaver, Porcupine, Silver Hill, Cariboo, 3A, and later, the Badger.[105] More discoveries of native silver and silver ore were made a few miles to the south, at Rabbit Mountain and at Whitefish Lake. But with the exception of the Beaver and East Silver Mountain mines, the owners of the mines in this new region were more interested in proving the value of the veins than in working them.

The developments at the Beaver Mine were impressive, given the general unwillingness of other companies to develop the mines in this region. The Beaver was discovered in 1882, during construction of the railway, but it was not opened to development until the railway was completed in 1883. Its owners, who were locals, sold it to American businessmen, who proceeded to open it to full-scale development.[106] A mill was erected on Silver Creek a few miles distant from the mine shaft. Supplied by the Fraser and Chalmers Company, the mill was equipped with what had become standard fare: a Blake Crusher, patented pulverizers for crushing and grinding, Frue Vanners (four), a Golden Gate Concentrator, and amalgamating pans. Within another two years, a Dodge Crusher for sampling purposes, two batteries of Fraser and Chalmers patent stamps (Gates), and a Krause separating table had been added.

The underground workforce, which numbered forty-five, likely drilled by hand methods since only one air drill, a Rand, was in use. The miners were paid between $2.00 and $2.50 a day, the labourers $1.50. A Copeland and Bacon hoisting engine operated at one of the shafts. The operators purchased their own diamond drill. The cost of boring with the diamond drill was seventy-five cents per foot, it taking one hour to drill down two feet. In addition to the extensive use of imported technology, one technological innovation is noted for this operation – Hooper's Improved Slow-Motion Vanner. A custom-built improvement over the Frue Vanner for treating the silver and gold ores of this region, which contained light, thin leaves of mineral, it was developed in 1890 by the mine superintendent, Thomas Hooper.[107]

The story of East Silver Mountain Mine was somewhat different from the Beaver's, for here the company of operators, who were English, decided to delay construction of a mill. They preferred instead to invest their money in underground development work and simply to stockpile the ore and await new developments in transportation which would make it profitable to dress and ship it. Ten thousand dollars were spent on an air compressor pipe, air drills (four Rands), a hoisting engine, pumps, boilers, and tools.[108] Unlike the Beaver Mine, however, the East Silver Mountain Mine simply did not prove exceptional enough to offset the high freight charges that plagued the mining operations in this inland area.

The Lake of the Woods Gold Region

Northwest of the height of land boundary of Ontario, the impossibility of obtaining secure title to land in the disputed territory, the frequent clashes with powerful timber interests in the Lake of the Woods section, and the lack of cheap, convenient transportation made it difficult to attract investment capital. Consequently, little in the way of immediate, serious mineral development occurred prior to 1890. There were a few gold-mining operations notwithstanding this inconvenience, as the following examples reveal, but the rebellious character of the ores and the frequent lack of reliable sources of water created serious problems for concentrating and smelting operations.

First, new interest was shown in the Jackfish Lake goldfield, which had been discovered about 1871, but not until 1873 was Treaty 3 successfully negotiated with Indians in the region to gain Canadian title to the Rainy River-Lake of the Woods country in which the goldfield was situated.[109] Peter McKellar, the local mining promoter, together with investors from the Ottawa area formed the Huronian Company to work the most promising orebody.[110] However, the Huronian Mine, indeed the entire goldfield, remained idle for a decade owing to lack of proper transportation and the absence of clear provincial title to the land.[111] Now, with the completion of the lakehead-Winnipeg section of the Canadian Pacific Railway in 1883, shafts were sunk at the old Huronian property. Under Peter McKellar as mine manager, ore was extracted and a ten-stamp mill, with amalgamation plates and three Frue Vanners, was erected, but too much of the gold was being lost using the limited gold-milling technology of the day. Besides which, the railway ran too far north of this area. The overall freight costs were so high that nothing but high-grade ore was profitable.[112]

Transportation was not such a serious problem in the Lake of the Woods gold region as at Jackfish Lake; here cargoes could be inexpensively freighted by water to the railway terminus at the old fur-trading centre, Rat Portage. The dependence on lake navigation, however, meant that mining operations were confined to whatever deposits were discovered along the shoreline and on the islands. Local prospectors, lumbermen, and others from Rat Portage, Winnipeg, and the state of Minnesota started up a handful of gold-mining operations — the Sultana, Keewatin, Argyle, Pine Portage, and Winnipeg Consolidated being the ones which were actually worked.[113] But for reasons already stated, these operations were marginal and underdeveloped in the 1880s. The gold occurred in such narrow veins, often only a few inches wide, that even highgrading was difficult. The most serious difficulty, however, lay with the milling end of the operations. Gold mills, complete with gravity stamps, Frue Vanners, and amalgamating pans, were erected at the most promising properties, but the operators quickly discovered what the operators of the Huronian Mine had previously: the existing gold-milling technology was inadequate for recovering the gold from these ores.[114] Only with the introduction of an entirely new separation process based on cyanidation would gold-mining in this region, and throughout the entire Shield territory, become profitable.

Nickel-Copper Mines of the Sudbury Basin, the 1880s

Construction of the transcontinental railway north of the Ottawa-Huron Tract accidentally uncovered the extraordinary nickeliferous copper pyrites of the Sudbury Basin in 1883. This was to become a great new, well-publicized mining region that would quickly overshadow the older mining districts of the province.[115] The discovery of copper ores by railway workers (the Murray Mine) touched off a rush of local prospectors and Ottawa Valley lumbermen to Sudbury Junction. They combed the region, purchasing mineral lands from the Crown at a dollar an acre, and for the ensuing decade the main activity was prospecting for workable deposits.[116] The Worthington and Frood Mines were discovered in 1884, the Stobie, Copper Cliff, and Creighton Mines in 1885, the Evans Mine in 1886, followed in 1887 by the Vermillion Mine. All these operations went on to become legendary in Canadian mining history.

Because the major sorts of mechanical equipment had already been introduced and were tested in North America's older mining regions and because the deposits were so enormous, the advance from hand work to mechanical production occurred quickly in the Sudbury region. The diamond drill, for example, was used extensively from the beginning as were compressed-air drills. The activity was mainly financed by a few heavily capitalized foreign companies of operators, who quickly consolidated the individual claims, imported miners and American metallurgical experts, and strove to develop appropriate, sophisticated new methods for recovering the copper and nickel from the ores of the entire Sudbury Basin.

The American-owned Canadian Copper Company began operations here in 1886, within a year of being organized by Samuel J. Ritchie.[117] Canada Copper quickly came to dominate operations in the region. It owned the best orebodies – the Copper Cliff, Stobie, and Evans mines, located close to the railway line and only a few miles from each other and from Sudbury Junction. The company's only potential competitor for the first few years was the celebrated Henry H. Vivian and Company, of Swansea, which had prospected the Murray

Mine for a few years under bond and only purchased it in 1889.[118]

Initially, only the copper in the ore was valued, but scientists elsewhere had recently discovered the strengthening properties of nickel when alloyed with steel. Canada Copper was the first to discover and, through the efforts of Ritchie, eventually to create an international demand for the nickel contents of the Sudbury ores. Much of the credit for the discovery was owed to the mine manager, Edward D. Peters, Jr., and the mine captain, F.F. Andrews, both of whom were internationally trained and highly experienced in copper-smelting. Peters, who was a prolific writer on the subject of modern methods of copper-smelting, charted the early developments at Sudbury in various published works and in his testimony before the Ontario Royal Commission on Mineral Resources.[119]

In the first years of operation, the company concentrated its efforts on Copper Cliff, where it also established a townsite and erected its smelters. Unlike with the more traditional mining practices, with mass-mining techniques the extractive end of the operation was conducted by only a few dozen men. Testing with a diamond drill had indicated massive beds at all three sites, so the work of "mining" consisted mainly of drifting and quarrying with Ingersoll air drills. Because it appeared to be the richest, the Copper Cliff Mine was driven the deepest; because it was deep, it was outfitted with a large double-cylinder hoisting engine and an automatic skip. In mass-mining the skills and experience of miners had little value; the miners therefore lost a considerable amount of their former status. Certainly, this was the case at the Copper Cliff Mine in 1889, where the operators of the hoisting engines received much higher wages than did the miners, who, next to the surface labourers, received the lowest pay ($1.75 per day) in camp.[120]

The first smelter was blown in at Copper Cliff in December 1888, followed by a second furnace in 1889.[121] The furnace chosen by Peters for Copper Cliff was the innovative new water-jacketed blast-furnace of the Herreshoff patent, with which Peters must have become familiar in his previous work at major copper-smelting operations in the United States. It seems that the company purchased the rights to the furnace in the United States, but the furnace itself was manufactured at Sherbrooke, Quebec; probably this arrangement was intended to avoid the heavy import duty on mining machinery. Ore from the Copper Cliff mine travelled in automatic skips to a rock house, where it was broken in a Blake Crusher, sized in revolving screens, and carried automatically outside to the open-air roasting yard, where huge piles of logs were covered with ore and ignited. There were thirty beds in the yard, each capable of roasting four to six hundred tons of ore which smouldered away for months on end. The fumes given off in the process destroyed the vegetation for miles. Despite the obvious hazards, roasting in this way was very cheap and effective; it drove off the sulphur and united the iron with oxygen so that the ore as taken from the heaps was largely self-fluxing. In fact, the ore from the Stobie Mine was so high in iron that it provided all the necessary flux for smelting with the richer but more rocky ores from the Copper Cliff and Evans Mines.

Within a few months of commencing to smelt, technological improvements were already being made; these included a "homemade" hydraulic jig for concentrating fine ores which were judged too rocky to be profitably smelted as well as a method for assaying nickel using electrochemical procedures.[122] These local improvements represented only the beginnings of developments, however. Since nickel was a rare and relatively unknown mineral, considerable investment and much expert knowledge would be required over the next decades to work out practical commercial uses for it. Out of this activity would eventually emerge one of the wealthiest mining companies in the world.[123]

Even without the presence of nickel in the ore, however, the future for mining at Sudbury would have been considerably brighter than for the earlier mining in the Upper Great Lakes district. The mineral occurences were in the forms of large masses and impregnations, and while the percentage of metal content was low, the quantity of ore was so enormous, even at depth, that the deposit was very easy to detect and work. Moreover, the area was already surveyed, with a railway and a supply centre, Sudbury Junction, in place prior to mineral development. The Canada Copper Company even managed to obtain highly favourable freight rates for bringing in coke from Cleveland and for shipping out ore.[124]

In the final analysis, transportation and mining technology had played relatively equal key roles in determining the pace of development of mining in the entire Upper Great Lakes district.

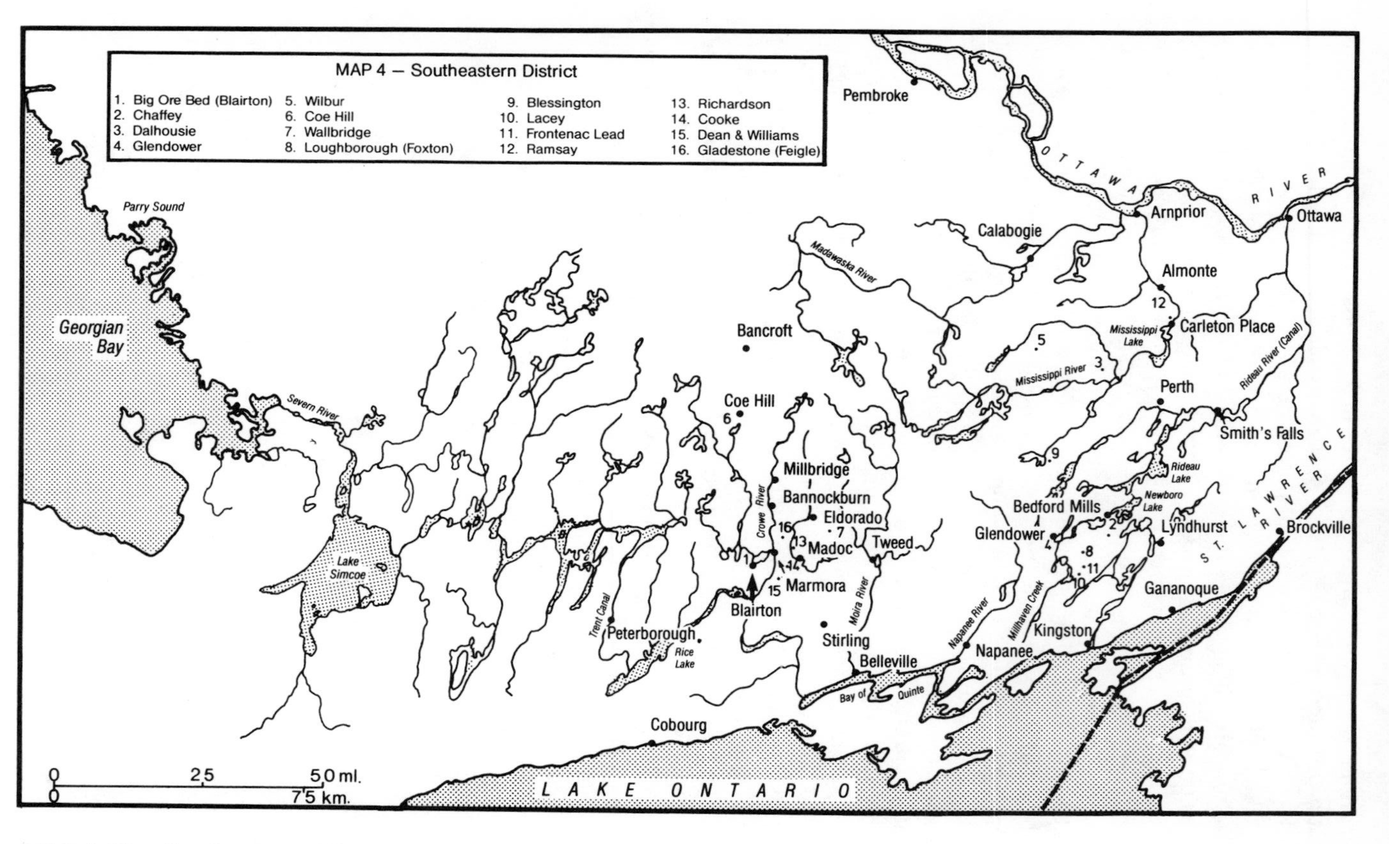

MAP 4. The Southeastern Mining District.

5

SOUTHEASTERN DISTRICT

The southern extension of the Precambrian Shield into eastern Ontario was opened to settlement fairly early in the province's history, but it proved of little agricultural value. To sustain early agricultural settlement, lumbering was important. However, by 1870 the prime timber in the district had been cut. Mining, on the other hand, by virtue both of its mystique and of its genuinely more permanent values, maintained commercial interest and evoked dreams of growth and prosperity. In the beginning, mining, like lumbering, provided seasonal work opportunities to supplement the largely marginal farming operations. The wide variety of minerals worked from hundreds of small sites included iron, which started up in northern Hastings County in 1820 and introduced settlement to the southwestern border of the south Shield region; apatite and mica, worked only on a small scale and exclusively in the Rideau corridor, which enjoyed the widest markets and ultimately proved the most commercially valuable; lead, which was marginal and shortlived; and hardrock gold, which caused a brief rush to the north Hastings County area in the mid-1860s and was responsible for the greatest efforts at experimentation with the new technology.

Because of its more favourable social and economic setting and the wider variety of minerals, mining in the district should have advanced beyond the levels of development and technological adaptation attained in the Upper Great Lakes. But only in a few rare cases did it even go beyond surface operations worked largely with muscle power. It is therefore important to examine because of the contrast it presents.

The district comprises a rugged upland of ancient igneous and metamorphic rocks. The Rideau waterway, which stretches between Ottawa and Kingston, marks its eastern fringe, while the western flank contains the Crowe Watershed and the Moira River, both emptying into Lake Ontario. This district contrasts sharply with the surrounding gentle lowlands. It is a region of rocky knobs, mostly of Precambrian age, covered with numerous lakes, shallow soils, and extensive forests. Iron ore, the most widely distributed of all minerals in the district, occurred in two forms: red hematite in an extensive bed situated in Renfrew, Lanark, Frontenac, Leeds, and Hastings counties; and magnetite, found in deposits among the Laurentian formations in the same counties, along with Peterborough and Haliburton. Lead was found in the iron-bearing area of north Hastings County and in even more promising deposits in Lanark and Frontenac counties in the vicinities of Carleton Place and Kingston. Gold was present in paying quantities only in the vicinity of Marmora and Madoc; apatite and mica lay in beds, deposits, or veins in the Laurentian formations, chiefly in the vicinities of Perth, where it appeared in fairly regular seams, and Kingston, where it occurred in irregular masses in granite. Overall, deposits of all minerals were small and often irregular at depth, and generally the ores were of poor grade.

The social and communications setting was clearly better than that of the Upper Great Lakes district. Lakes and rivers draining into the St. Lawrence waterway along the Precambrian contact zone linked the interior with international markets. Where each of these joined the St. Lawrence, a town or village had arisen which could function as a transfer point for men, capital, and equipment moving to the mining district in its hinterland and also for the products of the mines moving to distant markets. Along the western flank of the axis, Marmora and Madoc had sprung up before 1840 in response to local iron-mining interests. These villages were not solely dependent on mining, however, for they also served lumbering operations and the agricultural activities of nearby fertile areas. On the eastern margin along the Rideau system, villages such as Perth, Smith's Falls, and Bedford Mills had also been established earlier in the nineteenth century during the early military settlement and Rideau canal-building days.

By 1856, overland communications facilities were also available to improve access to the mining sections of the district. The Grand Trunk Railway was completed along the St. Lawrence from Montreal to Toronto, while the Cobourg and Peterborough Railway reached the Crowe Valley watershed at Rice Lake to serve the needs of lumbering and iron-ore mining. The latter railway was reorganized in 1868 as the Cobourg, Peterborough, and Marmora Railway, at which time a branch line was built from Trent Narrows to the Marmora mines. Additional local railways, such as the Brockville and Ottawa, which was completed in 1859 as a lumbering railway to Almonte via Smith's Falls (with a branch to Perth), and the Belleville and North Hastings, 1879, were built along the fringes of the district. In 1883, the Central Ontario was laid through Marmora, Eldorado, Bannockburn, and Millbridge to connect the iron-mining area of Coe Hill with Lake Ontario at Trenton. A year later, the Kingston and Pembroke line was completed from Kingston to Renfrew, also to accommodate iron-mining operations. At the end of the decade the Irondale, Bancroft, and Ottawa Railway was being constructed from Kinmount to Ottawa, to serve the iron-mining interests of Haliburton. Most of these lines were built with the help of government and mostly in response to mining activities, but those few which survived mainly did so because of the traffic from lumbering and farming.[1]

A system of colonization roads also began in the 1840s in the axis and greatly expanded during the 1850s and 1860s.[2] New roads and branches filled the gaps between the original routes of the previous decades. By means of these roads, canals, and the railway lines, the towns along the St. Lawrence were connected with the mining operations in the southeastern district to the north of them.

THE PREDOMINANCE OF IRON-MINING

Iron mines were the only operations that were widely extended throughout the district and continued for the entire period of study. They inspired little in the way of technological innovations or borrowings, yet they had a marked impact on the development of local transportation networks, especially rail lines. Before 1840 the little mining that took place in the district was confined to a few isolated experiments to smelt the iron ore collected from numerous surface deposits and market the iron locally.[3] Between 1820, the date of the opening of the historic Marmora Iron Furnace and its nearby ore bed, and the mid-1870s, a number of local and a few foreign capitalists attempted to overcome previous problems, only to succumb to new difficulties.[4] This pattern would characterize iron-mining activities, old and new, in the southeastern district until the late 1880s, when most operations, except for small-scale surface workings, ceased altogether.

The Marmora-Madoc Area, 1820-1875

The first serious attempt to exploit the fairly extensive deposits was at Marmora, where a charcoal-burning furnace was opened in 1820 by an Irishman named William Hayes to produce pig iron for the domestic market. The extent of Hayes' original grant was approximately three thousand acres.[5] Four years later the "Big Ore" bed (later renamed the Blairton Mine) was discovered and opened to development. The Marmora Furnace was located on the Crowe River, but because the Crowe was not navigable below Marmora, pig iron from the furnace had to be hauled thirty-two miles overland by a company-built road to Belleville. The Big Ore bed did furnish the dependable supply of ore needed to make smelting profitable, though the labour force, with its "wicked combinations" and occasional drunkenness, was considered troublesome by Hayes and his backers.[6] The latter included the prominent Montreal entrepreneur Peter McGill, who acquired the property outright in 1824 then disposed of it in

Plate 10. Watercolour by Susanna Moodie, "The First Mine in Ontario at Marmora." *This watercolour by the well-known author of* Roughing it in the Bush *was likely produced sometime between the years 1832 and 1840, when she lived in the southeastern district. It probably depicts the "Big Ore" iron mine (later renamed the Blairton Mine), opened in 1824 and worked off and on into this present century.*

1830 while retaining a financial interest.[7] By then transportation had proved too costly and difficult for the operation of this furnace to continue. Still, there had been some spreading effects. The iron works was responsible for establishing a village, in effect an iron plantation, along the lines of the ones near Trois Rivières, Quebec, and the Lake Champlain, New York, region and for the laying of the only road that stretched from the settled front of Hastings County through empty townships to its northern, Shield region.[8]

In 1837 another group of entrepreneurs erected a second charcoal furnace at Madoc to smelt ore from the recently discovered Seymour Mine.[9] Uriah Seymour applied experience gained with his New York State foundry and smelter, located at Wolcott, Wayne County, to build an experimental and moderately successful local enterprise. He made certain adaptations to existing iron-making technology to accommodate local conditions. For one thing, he experimented with different materials for flux and settled on local sandy loam. For another, when the supply of local charcoal ran out, he replaced charcoal with cordwood as fuel.[10] Two tons of pig iron per day were produced here for a predominantly local market. His venture, however, was shortlived; the death of a partner, personal financial difficulties, and a drop in the price paid for pig iron all led to the furnace's closure in the mid-1840s. The ore bed itself continued to be worked off and on for several more decades. The mine and the furnace resulted in the establishment of the town of Madoc and in the construction of a second road – this time to link Madoc with the Marmora-Belleville road at Stirling.[11]

During the 1840s the veteran iron-master Joseph Van Norman made an unsuccessful attempt to rejuvenate the Marmora Furnace and Big Ore bed.[12] Van Norman had already undertaken a few innovations in hot-blast technology and achieved some success in iron-making at Normandale and elsewhere in the province.[13] In 1847 he refit the Marmora furnaces, installed new machinery, ovens, and blowing apparatus, cut cordwood and produced a stock of charcoal for fuel; the furnace was blown in the new year. To overcome the obstacles presented by the thirty-two-mile wagon route to Belleville, Van Norman developed a new route. Pig iron was transported across Crowe Lake to the mine; from there it travelled nine miles via a company-built road to Healy's Falls on the Trent River. A steamer carried the pig iron to Rice Lake, where it was loaded onto carts and hauled approximately fifteen miles to Cobourg.[14] The pig iron fetched $30 to $35 per ton locally, which gave Van Norman a reasonable profit. Yet, almost immediately the furnace faced serious competition from cheaper English pig iron, which was being supplied to the Canadian market as backhaul cargo at $16 per ton. The difficulties and expenses of handling and freighting such a bulky, low-value commodity from the Marmora area proved too great for continued development and Van Norman closed down the works.

Other attempts to rejuvenate the Marmora furnace were made in the 1850s, first by the Belleville-based Marmora Iron Works Company and then in 1856 by

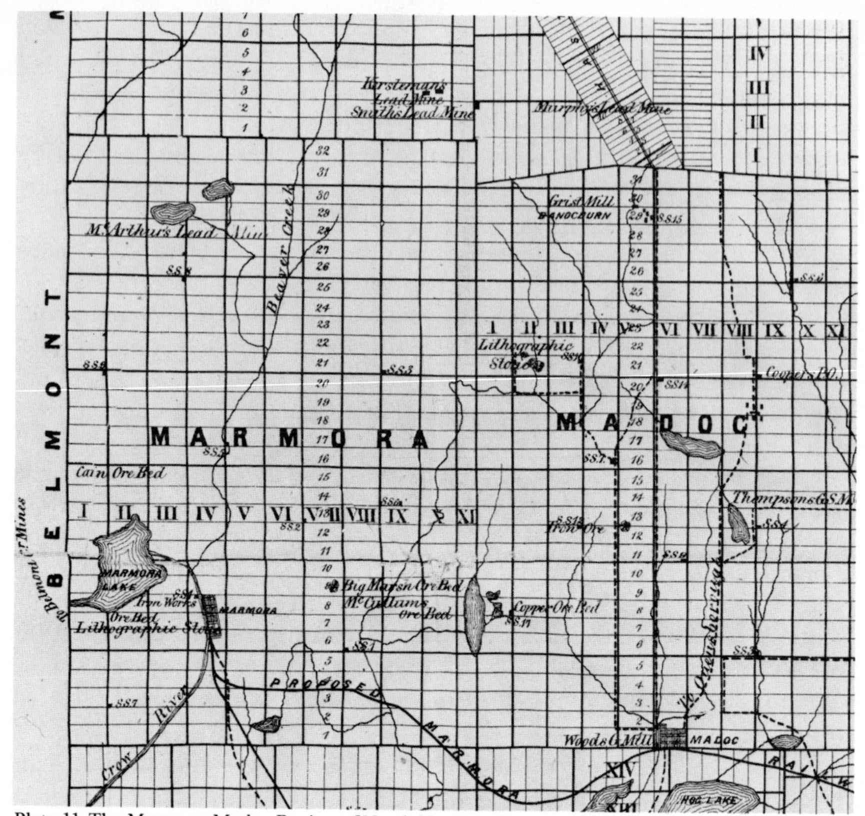

Plate 11. The Marmora-Madoc Region of North Hastings County, 1865. *The Hastings County Atlas for 1865 illustrated how closely linked were the small scale mining operations and the settlements in this marginal agricultural region. Hardrock gold deposits were discovered here a year later.*

an English company with Montreal and Quebec City capital.[15] The furnace was rebuilt for the English owners by Vernon Smith, whose experience with iron-making had been gained in Woodstock, New Brunswick, where a hot-blast furnace had operated briefly in 1848.[16] Fuel with which to adapt the new hot-blast techniques was a problem, and the machinery was in constant need of repair. The old cold-blast technique was reintroduced, but the high general operating and transportation costs for this location and long-distance supervision by the English management made the fledgling operation impractical.[17] The Marmora furnace was shut down once again.

The iron-mining industry finally came into its own, if only temporarily, in the late 1860s, thanks almost exclusively to American capital and markets that viewed the Canadian deposits as handy sources of raw material for their own iron and steel industry. The owners of the Marmora operation attempted to revive operations by amalgamating with the Cobourg and Peterborough Railway and using capital from Rochester and Pittsburgh.[18] The resulting new company – the Cobourg, Peterborough, and Marmora Railway and Mining Company – reopened the Big Ore bed, renaming it the Blairton Mine, and directed its ore-mining efforts towards this site, leaving the furnace closed for the time being. The operation reached its peak in the period 1868-73, when it was reported to be the largest iron-ore producer in the Dominion, although technologically it was quite unsophisticated.[19]

The main operation was an open pit which eventually reached one acre in surface dimension and a maximum depth of 125 feet. Its scale and depth required steampowered drills and hoists and a workforce, at its peak in 1869, of three hundred miners, some of whom were recruited in England. The village of Blairton mushroomed. The unprocessed iron enjoyed an assured market – the bulk of it was consigned to the company's smelters in Pittsburgh.[20] The route over which the ore travelled, however, was as complicated as in Van Norman's days and probably very costly: the company railway transported the ore from the mine eight miles to the Narrows on the Trent River, whence it travelled by scow to a Rice Lake transshipment point for transfer to rail to Cobourg. Then at Cobourg it was shipped across Lake Ontario to the United States side and thence by a third rail transfer to Pittsburgh.

A last attempt to reopen the old Marmora furnace to process the Blairton ores was made in 1875. This time the owners experimented using a cheap product of petroleum tar and sawdust being developed in the Ontario petroleum district to fuel the blast, but the experiment was not a success.[21] Because the local iron ore became depleted, the venture failed for the final time. Blairton became a ghost town in 1875.[22]

The Rideau Area, 1850-1890

Iron-ore operations shifted to the new deposits uncovered in the Rideau area, but these were on a much smaller scale than those at Blairton. While several dozen small iron mines were opened in the first few years of the 1870s, the work was still very much at the preliminary testing stage, involving some surface mining. Only a few operations actually advanced to sinking and drifting work, and these were exclusively deposits close to the route of the Rideau Canal in Lanark and Leeds Counties as the following examples reveal. An excellent case in point was the Chaffey Mine, one of several mining properties owned by this pioneer family.[23] The location had been worked in a small way since 1858, when it produced six thousand tons, all of which were shipped to Pittsburgh. Because of its favourable location on an island in Newboro Lake, the cost of mining was relatively low, which may account for the mine's longevity. It was a simple affair employing little in the way of mechanical power. A horsepowered crane raised the broken ore from the openings onto a barge bound for

Kingston, forty-four miles distant via the Rideau system. The nearby Matthews Mine has a comparable history.[24] It shipped to a Cleveland market.

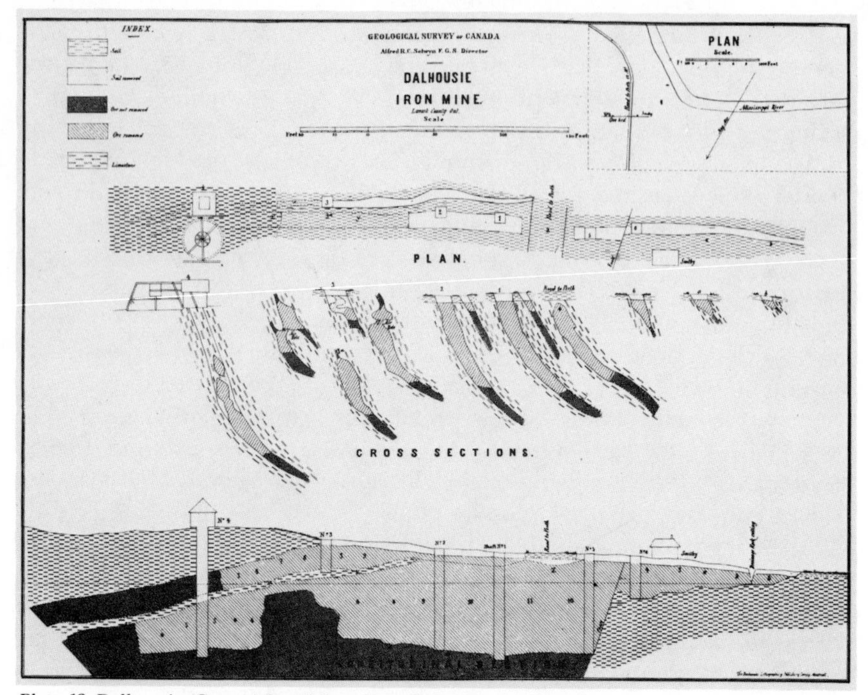

Plate 12. Dalhousie (Cowan) Iron Mine, Lanark County, in Plan, Cross Section, and Longitudinal Section (1872-73). *These drawings by members of the Geological Survey of Canada indicate the shallow-lying nature of the mineral deposits in the southeastern district. Operations here had to be conducted on a small, simple scale.*

The most successful iron-ore operation in the Rideau section of the district before 1873 was the Dalhousie, or Playfair, Mine located fifteen miles northwest of Perth. It also was a modest affair (see Plate 12). Opened in 1867 by Alex Cowan of Brockville, it consisted of surface work and five unconnected shafts by 1871, two of them to depths of seventy-five feet.[25] Each shaft contained a specular ore of No. 1 Bessemer quality, almost free of phosphorus, sulphur, and titanium. The ore was drawn to Perth only during the winter months, at a cost of $1.10 per ton. It was shipped from Perth to Brockville by rail and then to the smelter at Cleveland by water.[26] The previous year the shafts had filled with rainwater to the point where the horse engines were unable to keep them dry, and because the operation was so marginal, the shafts were simply abandoned.[27] One of the earlier shafts was reopened and deepened to ninety-four feet, and reasonable quantities of ore were extracted. It was hauled to Perth, where it travelled by barge seven miles to the Rideau, thence to Kingston and up Lakes

Ontario and Erie to the Cleveland market.[28] No more than twenty-five men were employed at any one time to work this mine.

Clearly, even the best iron operations were marginal. With the poor price and market situation that prevailed after 1873, iron-ore mining entered a decline, and only operations along the waterways and rail lines continued to be developed. Deposits actively worked in the Rideau section included half a dozen locations owned by the American-backed Kingston and Pembroke Iron Mining Company and operated on small scales by contract or by individual lease.[29]

The greatest success story in adapting new mining technology at these iron mines was the transformation of the Glendower mine into a full-scale operation by the Kingston and Pembroke Iron Mining Company. After being worked irregularly from the 1860s by a local miner, several locations in the vicinity of Bedford Mills were leased in 1873 to the Glendower Company of Elmira, New York, to be operated on a royalty basis. These locations likely belonged to the Tett and Chaffey families of Bedford Mills, who, in addition to owning sawmills, general stores, shipyards, and a forwarding company, developed local iron, apatite, and mica mines.[30]

The royalty was twenty cents per ton extracted, and twelve thousand tons were shipped out in the period from 1873 to 1880.[31] The ore was to supply the owners' blast furnace in Elmira, and it had to be at least 60 per cent in iron content to cover the royalty and the high costs of hauling it by wagon or sleigh the four miles to Bedford Station, but it proved too poor in quality.[32] Subsequently, in 1880, the mineral rights were purchased by local entrepreneurs, the Folger brothers, who expended $40,000 on its development but to no avail.[33] From the Folgers the entire property passed, in 1882, into the hands of Cleveland and Zanesville, Ohio capitalists, who operated it as the Zanesville Mining Company. Whereas the Glendower Company had been capitalized at $50,000, the Zanesville Company was capitalized at five times that amount. At this point, therefore, the property underwent what amounted to full-scale development in this district. The large plant introduced included a compressor works ("National") driven by a 100-horsepower engine, a battery of fifteen Ingersoll compressed-air drills, two Cameron pumps, and compressed-air hoisting arrangements for one shaft.[34] Practical compressed-air drills had been available for less than a decade, so their adoption at this operation was timely. It probably was the first of few occasions when they were used in mining in this district.

About three hundred men were employed and three thousand tons of ore were extracted from shafts which reached three hundred feet during the first year of the mine's operation by Zanesville. The operators were sufficiently committed to development, besides, to construct a four-mile spur rail line in 1884 to link with the Kingston and Pembroke at Bedford Station, where it could proceed to Kingston and the Cleveland market. The spur rail line cost the company $25,000 to construct.[35] Within a year, however, the ore began to prove even leaner than in the Glendower days; below 120 feet the miners encountered ore with less than the now-requisite 50 per cent iron content and high in sulphur.[36] These discoveries coincided with a depression in the iron market by mid-decade; together they caused production to be halted and forced the Zanesville Company to merge with the recently formed Kingston and Pembroke Iron Mining Company in 1887.[37] The latter firm resumed work on the location and drove a shaft to four hundred feet, where they found higher-grade, sulphur-free ore. That ore quickly ran out, however, bringing an abrupt final close at this once-promising site.

The neighbouring Wilbur Iron Mine followed a similar though less spectacular course.[38] By the early 1880s the shafts reached a depth of two hundred feet, with a 350-foot drift to follow the vein, which dipped off at an angle of 45 degrees. The operation was taken over by the Kingston and Pembroke Iron Mining Company in 1886. Compressed air was used at this mine as well as a recently developed "Legerwood" hoist imported from the United States. The ore, which ran at 55 per cent iron content, was shipped from the mine via the Kingston and Pembroke Railway to Kingston, and from there it was bound for a Bessemer steel plant in Pittsburgh. When members of the Ontario Royal Commission on Mineral Resources visited the site in 1888, only ore at the surface was being worked. Still, exploration was proceeding with a diamond drill, and it was then the sole iron mine surviving along the line of the Kingston and Pembroke. All the others were undeveloped or idle and flooded.

The Renfrew County Region, 1880s

In the decade of the 1880s iron-ore operations experienced still another shift of locale, this time to the northern parts of the district being opened to settlement. Once the northern terminus of the Kingston and Pembroke line reached the Madawaska River area, new iron deposits were discovered and worked. The most ambitious was that of the Calabogie Mining company, backed by Perth and Toronto businessmen, along with B. W. Folger of Kingston, who was superintendent of the Kingston and Pembroke Railway and a principal in several iron and apatite properties.[39] The mine opened in the 1881-82 season, and a two-mile road was immediately constructed at a cost of $1,000 to join with the Kingston and Pembroke line.[40] This operation was fairly characteristic of iron-ore works in this northern section of the district. Hand drills and black powder and dynamite loosened the ore from the pit that extended four hundred feet along the vein. Because the pit opened to a depth of one hundred feet, steam pumps were employed to empty it of water. The twenty-five-horsepower engine that operated the Cameron pumps also powered a hoisting engine. One innovation is recorded: this operation employed locally improved self-dumping buckets.[41] Progress towards opening the location was typically slow, and the first ore did not reach Kingston until 1883. The limited production plus the adverse market conditions forced the company to halt work at the location for three years. When the company renewed exploration and development in 1886, a new steam drill was in use.[42] The value of the entire steam plant at this point was $5,000, and twenty to twenty-five men were employed. Promising veins of Bessemer ore were soon discovered, whereupon the company followed the example of others by promptly leasing the location to the Kingston and Pembroke Iron Mining Company, which worked it for one year until the deposit showed signs of depletion. When the Royal Commission party reached the site in 1888, however, the operation stood idle and the pit was filled with water.[43]

Another mine of interest in the Renfrew County region was a small iron-pyrite operation near the town of Almonte. Iron pyrites are a source of sulphuric acid, and the associated iron is of little commercial value, but since the mineral is mined as iron ore, the operations are described here. This mine, the White Birch, was opened in 1882 by William Wylie, a Perth woollen textile manufacturer and mining entrepreneur. The deposits were quite extensive, but the mining operation itself was simple, drilling and pumping in the single, 140-foot shaft being wholly conducted by hand by one man, Charles Clymo, a Cornish miner.[44] The entire output of iron pyrites was bound for the Brockville Sulphuric Acid Works in Elizabethtown, which belonged to the Brockville Chemical and Superphosphate Company. The ore was hauled by road to the Almonte terminal at a cost of $1.50 to $2.50 per ton and thence via the Brockville and Ottawa Railway. The Elizabethtown works, which had been opened in 1874, previously had purchased its pyrites exclusively from a short-lived pyrite mine located close to the works.

Coe Hill, 1880s

In the western section of the district, iron-mining in the 1880s was also concentrated around local rail lines. In the older Hastings iron region, the Central Ontario Railway had been constructed in the early 1880s expressly to connect the iron region with the St. Lawrence River, the purpose being to freight iron ore from the new locations being developed in the Coe Hill area.[45] The leading figure in this venture was Samuel J. Ritchie, of Akron, Ohio, who was mentioned previously in connection with the Sudbury developments. Ritchie organized two companies – the Cleveland Mining Company and the Anglo-American Iron Company – that acquired vast tracts of mineral properties in the area transversed by the line.[46]

The region was dominated by a single mine – Coe Hill.[47] The location, discovered by William Coe of Madoc, started in a modest way, but in swift increments, it introduced steam technology on a large scale to the area. Side cuts were made in the hill and hoisting at first was by horse derrick, soon replaced by a small, twenty-five-horsepower steam engine. The principal surface excavation reached one thousand feet in length, and ranged from twenty to thirty feet in width. Soon shafts were sunk, and tunnelling proceeded on several levels. Now it became a full-scale operation. Steam rather than compressed air was employed because this, like most others in the district, was a surface operation. A large plant with a 275-horsepower engine was imported from the Globe Iron Works in Cleveland, and a complete townsite was built to accommodate twenty-seven families.

Just when it was beginning to look as though this was to become the long-sought permanent mining region, the depressed state of the iron trade and the serious competition offered by the large and uniformly rich Lake Superior iron fields in 1883 forced the company to test the entire 1,500-acre Coe Hill area thoroughly with a diamond drill before investing further. As a result of these explorations and the possibilities afforded by the Central Ontario Railway, the company opened numerous new mines – the Arthur, Chandos, Cleveland, Tudor, and Orton. But none of them ever amounted to much more than surface workings. A short spur line to join Coe Hill with the Central Ontario Railway was constructed, and five thousand tons of iron ore were shipped via that railway to Cleveland. By 1887, however, production at the minesite and plans for developing others had ceased owing to the high sulphur contents and unbeatable competition from the Lake Superior range.[48]

There was an interesting outcome of the building of the Central Ontario Railway and Samuel J. Ritchie's involvement in the undertaking. When the northern Hastings iron field proved unproductive, Ritchie and his associates found themselves with little freight. Fortunately for them, the downturn coincided with the discovery of the Sudbury nickel-copper field to the north, which led Ritchie to become a prime mover in establishing the Ontario nickel industry.[49]

The use of the diamond exploration drill at Coe Hill had a positive impact on other mines struggling in the area. At the Wallbridge Mine located four miles north of Madoc, for example, where small amounts of ore had been mined even in prehistoric times, serious development work occurred only in 1881, when this entire iron belt was explored with a diamond drill and developed.[50] The Wallbridge Mine deposit consisted of a large mass of hematite ore in dolomite, with no defined walls. The location was leased to the Bethlehem Iron Works Company, Bethlehem, Pennsylvania. The company produced one thousand tons of ore between 1881 and 1883, by which time the orebody had been depleted. Bethlehem had invested a modest amount in a small steam operation, which cost $3,500, and constructed a 5,400-foot spur line at a cost of $2,500 to connect the site with the Belleville and North Hastings Railway.[51]

The use of the diamond drill was exceptional in this mining district; likely it was employed because the iron from this district enjoyed, at least temporarily, a relatively sure market in the United States, so the cost of exploring with the diamond drill had a chance of being recovered from profits. The commonest cause of failure was the mode in which iron ores in this district (and in most of the Canadian Shield) were formed. They occurred in stocks, or limited masses of ore, lenticular or irregular in form, inclined or horizontal in position, and presenting no clear surface indication as to quantities of ore present. Also there was hematite, not detectable with a compass. E.J. Chapman investigated the characteristics of iron-ore deposits in this region and in the Precambrian Shield generally and recommended that the diamond-tipped exploration drill was required to test the character of deposits throughout their entire mass.[52]

Farther north still, in Irondale, the Bancroft and Ottawa Railway was built in the 1880s as a private line to serve the iron-mining interests in that section. Five new locations showed promise, some of which the railway company leased.[53] An American firm, Parry and Mills, undertook construction of a blast furnace near Parry Sound, but the venture failed because of a shortage of capital.[54]

It is clear that lack of adequate transportation was not to blame for the failure of iron-mining in this district. Canals and railways bringing access to the iron deposits were plentiful. The difficulties lay with the small size of most deposits, the poor quality and inappropriate type of most of the ores, the instability of prices, and, by the early 1880s, new American sources of supply.

DESULTORY MINING OF APATITE AND MICA

The second most active mining operations were in apatite, or phosphate of lime, and mica. None of the operations employed much in the way of machinery or took advantage of rail transportation: the deposits were located too close to the surface and involved minerals of such low value that costs – especially transportation – had to be kept at a minimum. Apatite- and mica-mining operations did not start in earnest until after Confederation, and in every case they were confined to a triangular area extending approximately fifty miles from Perth to Kingston in the vicinity of the Rideau Waterway.[55]

The most active period for the industry was in the late 1860s and early 1870s, mostly centred on the town of Perth. The first commercial shipment of apatite in Canada was from North Burgess, in 1870. For a time the production of the Perth area was marketed mainly in Germany. In fact, when a German importer named Schultz withdrew from the region in 1874, three operations, including his own in North Burgess, were forced to close.[56] A few other apatite deposits were worked for English, Irish, and American markets. The output of the bulk of the remaining operations was contracted by various owners to Alex Cowan for his Brockville Superphosphate Works (1869). In the 1870s the production from most mines had to be hauled over roads to Perth, then by scow along the Rideau Waterway, to Kingston, whatever the final destination. The products of the phosphate mines of Bedford and Loughborough Townships, which belonged to the Tetts and Chaffeys, were shipped by their own barges.[57] In 1885, a second plant – The Standard Fertilizer and Chemical Company – commenced operations in Smith's Falls, providing a second, even closer, local market for apatite.

Because apatite was mined in soft, crumbling rock, in which shafts and tunnels would have had to be extensively timbered, most mining took place relatively inexpensively at the rich surface deposits. There were a few exceptions to this pattern, however. At Alex Cowan's own apatite location the shafts reached 56 and 48 feet; at McKinley's (worked for Ritchie and Jackson, Belfast, Ireland), a well-timbered shaft was sunk to 40 feet, from which 200-foot tunnels were driven on either side. Even more noteworthy was the Floerstein and Company (of England) Apatite Mine, with shafts of 135 and 70 feet. At the latter, apatite of good quality and in fairly free form was found. Here alone, because of its deep-running shafts, steam engines were used for hoisting and pumping.

The method and scale of other operations was much simpler than that of Floerstein. Usually veins of apatite from hundreds of exposures situated on a single few-hundred-acre property were worked during the summer months. Most of the openings were shallow pits or trenches worked by contract. The miners, or "muckers," were paid $1.00 per day. They concentrated their efforts on the mineral and tended to remove as little dead rock as possible. As a result, they were unwilling to work below a maximum depth of approximately thirty feet for fear of the untimbered walls collapsing on top of them. The miners followed the configuration of the vein until they became cramped for space, at which point they simply abandoned the opening and started again on a fresh outcrop. The material raised was usually hand-cobbed at the pit's mouth. Perhaps because this sorting process required more skill than the extraction, cobbers were paid higher wages, between $1.10 and $1.50 per day. The boys employed to bag apatite were paid 75¢ per day. In these first years, the cost of raising the apatite and preparing it for shipment ran at $3.00 to $5.00 per ton; the cost to ship it by water to market was approximately $2.00, and the price paid for apatite at Kingston was $9.00 to $12.00, so a small profit was possible.[58]

After 1875, apatite locations were worked for a time mainly on a royalty basis. The royalty was set at $1.00 to $2.00 per ton, varying according to the state of the roads and the distance from markets.[59] The mining of apatite levelled off considerably in the 1880s, by which time the mineral was being mined from several dozen locations.[60] Many apatite mines remained unchanged from the earlier period, despite the machine-powered equipment available on the market. A given location typically contained a large number of shallow pits of various sizes and shapes, many of them abandoned and full of water, with only a few points of active development. The equipment usually consisted of horsepowered derricks and hand drills or, occasionally, steam drills. Pits were worked downwards as long as the ratio of mineral to dead rock was between 1: 5 and 1:10. Because of the hand-cobbing at the pit's mouth, most apatite properties were littered with extensive dumps of waste rock. At larger operations the cobbing was done in crude sheds.

At a few apatite deposits there were attempts to increase production, but these did not significantly alter the previous scale or the costs of development. Work at one, the Foxton Mine near Kingston, for example, commenced in 1877 in the usual way as a shallow, open pit. Within a few years there were dozens of openings being worked, and by 1883 the original pit had been enlarged to a 115-foot deep quarry from which both apatite and mica were mined.[61] A steam drill and steampowered hoist were engaged by 1887. The drill and hoist required so little skill or strength to operate, however, that they were run mainly by local boys. Since the location was situated close to the Rideau Waterway and only five miles from Kingston, the apatite could be delivered there by water for only $1.25 per ton; the cost to ship by rail would have been $1.75.[62] The price paid at Kingston for the Foxton apatite was a considerable improvement – 20 to 50 per cent higher – over the price levels of the previous decade. The market for most of the Ontario apatite operations had shifted from Europe to Montreal by 1885. When members of the Ontario Royal Commission visited the apatite area in 1888, only the Foxton Mine and two other mines owned by Boyd Smith of Washington, D.C., called the Blessington Mines, were still in operation.

It was complete folly to advance to full-scale development of these apatite deposits, as the following case will demonstrate. The district's largest and technologically most sophisticated apatite operation was that of the Anglo-Canadian Company, organized in Liverpool, England, which acquired a number of apatite locations in the Ontario and Quebec apatite belt including in 1886 the small-scale North Burgess and Bedford townships locations belonging to Robert Adams, who continued to mine the properties under directions from Liverpool.[63] The Anglo-Canadian Company attempted to introduce deep-level mining with compressed-air drills – the one instance of their recorded use in apatite-mining in the district. The nature of the deposits and the relatively low market value of the mineral meant that the machine drills were impractical and uneconomic. For instance, a great deal of waste rock would have had to be removed from the pits simply to accommodate the equipment, and this would be a costly undertaking. Since the seams were very irregular and narrow, they were best worked in the old way, and the return to it also meant a return to the contract system, for the company found that the deposits (as many as two hundred small pits) were so scattered that it proved difficult and, ultimately, too expensive to supervise the men. After a promising period, apatite mining was effectively eliminated in the late 1880s in the southeastern district. This was due to the discovery of extensive, rich, easily mined apatite deposits in North Carolina and Florida. The advantage of these new fields lay with the largeness of the deposit and their location in the United States, where they were close to an enormous market.

Deposits of mica had been discovered in the 1860s in conjunction with apatite, and they were welcomed, as the sale of mica could partly offset the costs of apatite-mining.[64] Unfortunately, the most lucrative type of mica was not common in this district. What there was usually occurred within ten or twenty feet of the surface.[65] Consequently, as with apatite-mining, few mica operations reached any depth.

The low level of mica operations is described in Geological Survey reports for the years 1870 to 1873. The earliest operation in Burgess Township was that of the New York Mica Company. The value of its plant and machinery was declared as a mere $250 in 1869, the one year of its operation. Yet that plant reported an impressive four thousand tons of mica raised by the efforts of twenty-two men and two horses.[66] The only deposits to be worked during the ensuing two years were at the Baker Mine in the same township.[67] Baker was a resident of nearby Perth who managed the operation for an English firm until the mine was closed in 1872. At first the uncut mica had to be sent to Perth for finishing, which significantly added to production costs, but shortly thereafter a small finishing shed was erected at the summit of the hill from which mica was extracted. Following rough sorting, or "thumb-trimming," at the pit, the sheets were knife-trimmed, sized, and split by boys at tables using tobacco-cutter knives. As with apatite, the mica was extracted on a contract basis, and so much debris accumulated at the site that it proved difficult to interest new buyers in the properties.[68]

Once the mica industry came into its own, in the mid-1880s, it was dominated by a single operation – the Smith and Lacey Mine, discovered in 1880 by apatite prospectors. Though this was the largest and most productive mica mine in Canada at the time, it was, nevertheless, like the iron-ore and apatite operations, undercapitalized and labour-intensive.[69] The owner was the local firm of Isaiah Smith and Company of Sydenham, dealers in mica and phosphate. The shaft was sunk in 1884 and reached 130 feet by 1888.[70] Twelve men, a few horses, but no power machinery were employed at the site, while eight men worked in the finishing shed. Here, as at the few other mica mines in this section of the district, mica was cut into various sizes and shapes for stoves and a few other purposes. In all, about a hundred tons were produced per year for Canadian, American, English, French, and German markets. Only about two hundred pounds of good marketable mica could be salvaged per ton taken from the ground. The price varied considerably according to the size and quality of the finished crystals.[71] In some cases the refuse was ground and sold in Canada to be used in the manufacture of lubricators, but the price for this was $45.00 per ton, a mere fraction of that paid for mica in crystal form.

At the very time that apatite operations were dying for want of suitable markets, the mica operations that had been auxiliary to them enjoyed a revival through the opening up of new markets for the lower grade amber mica, in powdered form, that was so common in the district.[72] A reflection of this new activity was the patented invention of a Copetown gentleman, Thomas Head, for grinding mica.[73] In addition, the value of mica as an insulator was now recognized and being exploited by the growing electrical industry in the United States.[74] As a result, during the late 1880s old abandoned deposits were rejuvenated and mica-mining in southeastern Ontario continued as a paying proposition well into the twentieth century.

LEAD, 1866-1880

Lead operations in the district featured quite different experiences with technological borrowings and innovation than iron-ore, apatite-, and mica-mining because lead was a high-value mineral and occurred in forms that required full concentrating and smelting at the site. Lead ore was found in productive quantities only in small areas in the vicinities of Marmora, Kingston, and Carleton Place. In most cases the operations proved unproductive and relatively shortlived.

Lead brought into existence a half-dozen full-scale operations which entailed underground mining and smelting works patterned after those in England and the United States. But each operation failed, almost always as a result of the nature of the deposits, which tended to be highly deceptive. The veins yielded considerable amounts of lead ore near the surface – with veins varying from one to thirteen feet in width – but they were found to diminish quickly below the surface. The irregular nature and low quality of the lead deposits at first gave operators little incentive to experiment with new technology.

The first lead deposit was discovered in 1858 in north Hastings County, near Marmora.[75] The site of the original discovery was leased by its owners to a Boston concern, Lombard and Company, which began development of the property. The deposit, however, was exhausted within a year. Other discoveries there and in the adjacent township of Lake followed, but no real developments took place. Only with the discovery in the mid-1860s of richer deposits of lead ore in Frontenac and Lanark counties did serious development work in lead-mining begin.

The Frontenac Lead Mine in Loughborough Township, only fifteen miles from Kingston, was based on a much more extensive deposit, and consequently its discovery in the mid-1860s attracted the attention of geologists and entrepreneurs anxious to learn about promising mineral deposits in the district.[76] Anticipating future developments, the operators immediately sank test shafts and erected the usual buildings for this type of mining operation – blacksmith's and carpenter's shop; a shed containing a stable, granary, and hay loft; a shaft house; a powder magazine; an ore shed; and a boarding house. Dr. J.W. Dawson, of McGill University, Charles Robb, then head of the Nova Scotia Geological Survey, and E.J. Chapman, the consulting engineer and professor of geology and mineralogy at University College, Toronto, examined the property and mistakenly recommended its full-scale development. Dawson even declared it to be "the most important deposit of lead in Canada or neighbouring parts of the United States."[77] He pointed out that operators at the Frontenac Mine could benefit from the knowledge derived earlier at a similar lead deposit being worked at Rossie, New York. Robb and Chapman, who analysed the ore body and its metal content, reported that it showed enough promise to warrant serious development work. The shaft had been sunk to forty feet, and the vein was found to be thirteen and a half feet thick at that depth. Robb's optimism was clear:

> this vein alone, both in extent, richness and the unusual facilities it offers for working is sufficiently important to justify a liberal outlay of capital in systematic mining and in the erection of permanent and substantial machinery for draining the mine, and for crushing, dressing and smelting the ore.[78]

His advice was followed, though in the end it proved unsound.[79] Each of these scientifically trained consultants was misled by the results of preliminary testing.

While no records describing the type of equipment installed have been found to date, the smelter was likely an American open hearth, as recommended by E. J. Chapman in his report on the mine.[80] A fair amount of metal would have been lost if the American hearth was employed, but Chapman considered that this loss would be more than offset by its efficiency in using the local pine for fuel, particularly if a practical concentrating mill that properly treated the ores prior to their reduction was introduced. The equipment used did not function well, however, and activity at the site was suspended for five years. In 1874, the discovery of other productive lead deposits in Lanark and Leeds counties led to renewed interest in the Frontenac Mine property. Further testing was undertaken, and on the strength of the positive results the property was leased for ten years to a Mr. F. Stockwell to be worked on a royalty basis. Stockwell promptly organized a company in England.[81] Using three American hearths, the deposit was worked at intervals until 1881, when it shut down for the time being. It had been poorly managed, but more important still, the quality of the ore did not warrant its continued development given the technology of the day.

A similar fate awaited the two other lead-mining operations that started up in 1874. One, called the Ramsay, was located near Carleton Place. In fact, it had been discovered about 1857 when a Montreal-based company was incorporated to develop it.[82] In 1862 the company's capital was doubled, and its headquarters was moved to London, England. The mine was worked on a quite modest scale by hand.[83] Most of the ore was secured from open cuts along the course of the vein, rolled, and calcined in an open hearth.[84] The Ramsay Mine was abandoned within two years because what had appeared to be a magnificent deposit on the surface proved to be too shallow and irregular to be worth the expense of developing.[85] Operations also commenced on a nearby Lansdowne Township property, but work there had also ceased by 1876. An English company of operators, the Canadian Lead Mining Company, had insisted on undertaking full-scale development; but despite the extensive use of steampowered equipment and the installation of a smelter, the operation had to be suspended after two years because of the low quality of the ores.[86]

GOLD, 1866-1890

Gold operations experienced the same disappointing trend as lead. Gold, however, was a glamorous metal with a sure market as a monetary metal. Thus, a few companies undertook serious attempts at full-scale development, which entailed the introduction of new technology and a number of local inventions.

Gold-bearing quartz was found in productive quantities only in a small area between Marmora and Madoc. In 1866, the search for copper by local prospectors, led by a farmer by the name of Powell, led to the accidental discovery of gold in paying quantities on a Madoc Township farm belonging to James Richardson. Henry Vennor, who was working in the area on assignment for the Geological Survey, identified the metal, but he wisely cautioned in his report that it did not appear to be a true vein, that the gold might be confined to a localized region, and that further examination was necessary.[87] Few people paid attention to his advice.

The deposit uncovered at the Richardson Mine was indeed a uniquely rich occurrence.[88] Despite Vennor's warning, the news of the presence of the coveted metal – especially in such a relatively accessible mining territory – touched off a frenzied gold rush by several thousand gold-seekers to the north Hastings area.[89] Initially, this rush caused such a stir that the local officials ordered two dozen mounted policemen to patrol the stretch of road between Belleville and Madoc. There were long-term consequences as well, such as the expansion of the existing town of Madoc and the establishment of a village – Eldorado – at the site of the first discovery. The entire area was thoroughly combed for further occurrences. More gold was discovered, but in disappointingly small amounts and in a form that almost defied treatment by known methods – the same problem that had plagued gold-mining operations elsewhere in the Shield in the Upper Great Lakes district. The truly difficult character of gold quartz occurrences was still not properly appreciated, and a number of gold-mining companies risked their capital before adequately testing the locations. Having done so, some companies continued seeking new ways to treat the difficult gold ores of the region and even to recover by-products in the process. In all, eight inventions in the treatment of gold quartz were patented between 1867 and 1872 by locals.[90] As a result, although gold-mining locations were confined to only a few townships around Madoc and Marmora, they proved the most experimental and technologically innovative of the mining ventures in the southeastern district.

The gold deposits were located adjacent to what Survey officer Henry Vennor described as "the western ferriferous belt," which he interpreted as the strongest preliminary indicator of probable sites for discoveries of gold.[91] The entire southeastern Ontario gold belt, in fact, was much more extensive. It stretched from the township of Belmont in Peterborough County eastward seventy miles across Hastings, Addington, and Frontenac counties into Lanark.[92] Nowhere, however, would gold be found in a form that would repay its treatment.

This fact was not known during the gold-rush days, as the unique diary of one adventurer reveals.[93] Thomas Nightengale was one of those struck by gold fever in 1867. He was a Madoc-area farmer who also managed to hunt and trap, haul timber, and perform odd jobs for other farmers and mine operators. He opened a small quartz vein on his family's property and with the help of friends worked it by hand during his slack time. He sank a shaft with hand drills and black powder; stoped ore by the ancient technique of firequenching to split the rock; and raised ore on a home-made windlass. It took a full day to chop enough wood for a fire and to empty the shaft of a day's accumulation of water. It took another full day to put in three shots, and yet another to haul out the ore which was loosened by the blast. When in a year enough ore had accumulated at the surface, he spent a summer putting in a small, primitive ten-stamp mill and raw amalgamation works to concentrate it and ore from neighbouring gold-mining operations. His neighbours donated the construction materials for the mill and engine house, which they also helped build. In his diary, Nightengale referred to this construction activity as a "bee."

Many accidents occurred as a result of his inexperience: a blast destroyed the windlass; freezing temperatures caused the boiler to explode on several occasions. His carelessness with an axe caused him to injure his leg; and his unfamiliarity with milling machinery resulted in his arm being caught in the cams of the stamp mill. And Nightengale was not alone in his misfortune. Over at Cook's Mine, the inexperience of workers there caused one man to be killed and another to have his arm blown off in a blast. Nightengale's milling machinery constantly broke down, the amalgamation pan leaked, and the crucible cracked, all of which caused delays and added to production expenses. To add to his problems were the tremendous physical discomforts he experienced. He worked long, strenuous hours in a powder-filled shaft which caused him headaches, in bitterly cold winter temperatures, and damp, mosquito-infested summer days. In the mill he experimented with a variety of methods of treating gold quartz. Those experiments included a process of separating by means of forced filtering through a constantly agitated mercurial bath, patented by a local millwright, Asa Long.[94] But too much quicksilver was lost using this process. Nightengale's small mining operation lasted from 1867 to 1871, but his labour and sacrifice were never rewarded.

By the time the gold rush had subsided in 1871, only seven localities, all of them in the Moira River Valley, had been steadily worked. Similar deposits known to exist in nearby Madoc Township were so hidden beneath horizontal limestone strata that they were difficult to locate or mine economically.[95] The development pattern was the same in each case. First, the location was cleared of bush and a road was built to connect with the main road between the villages of Marmora and Madoc. Then a few openings were made and shafts were sunk to shallow depths – usually less than fifty feet – along the most promising veins. From these openings, several tons of ore were raised and piled to await crushing and treatment in one of the various custom mills springing up at Madoc and Eldorado.

Mining operations like these and Nightengale's did not stand a chance in this district. The gold occurrences were mainly refractory, with high arsenical contents, which meant they required roasting to drive off the sulphides and soften the metal and special amalgamation procedures to overcome the adverse effects of arsenic on mercury, an essential ingredient in amalgamation. These procedures were too costly as well as unworkable in the gold-mining operations of this district.

Dean and Williams Mine

Technological experimentation with gold recovery in the Madoc-Marmora area reached its peak in the decade of the 1870s. The most advanced such operation was the Dean and Williams Mine.[96] Work was begun at the site in 1869 for a Toronto ownership under Dr. J.D.R. Williams as mine superintendent. With a workforce of thirty-six men and six horses, the site was cleared and a shaft sunk to a depth of seventy feet. The introduction of new technology proceeded in the usual manner – incrementally. Williams employed an old standard five-stamp battery with circular rotating heads, purchased from a mine in nearby Frontenac County. The stamp mill was powered by a small, 15-horsepower steam engine. Williams decided that the ores should be calcined in an open kiln prior to concentration – a traditional and very slow operation, four days being required to roast one ton of ore thoroughly. Yet the dramatic increase in yield from this treatment, crude as it was, led Williams to replace the old stamp mill with a newly patented quartz crusher.

Williams had wanted only to increase the production capacity of the mill, and it seems that he would have been content with simply acquiring used stamps secondhand from a nearby operation. A visit from a Windsor patent agent, however, caused him to undertake a different avenue. The agency, Macdonell and Foster, which was promoting a new crushing apparatus invented by I.W. Forbes, contracted with Williams for a trial of the machine. Williams had nothing to lose and possibly something to gain from the deal. The battery was not a success, and Williams had to return to his original plan to purchase a secondhand battery of stamps.[97] However, the story of the trial of this invention and the reasons for its failure, which have been preserved in a company letterbook, provide valuable insights into the problems of employing new mining technology for both inventors and mine operators.[98]

As described earlier, the Windsor firm of patent agents holding the Canadian rights to the Forbes Automatic Steam Quartz Battery were trying to break into the Canadian market by having the invention successfully tried at an accessible, known gold-quartz mine. The Dean and Williams mine was visited in 1870, and it was selected because it was close to the Grand Trunk line, with a good gravel road to the mine.[99] The special feature of the stamps was the elasticity of the mortar, which permitted greater amounts of force to be applied to the crushing operation without causing a proportionate amount of wear to the main parts.[100] A special, confidential proposal was made to Williams. He was offered two batteries at the cost price of $1,500 in gold, an amount which also included the costs of freight, customs, and installation. The mine was to provide the timber foundation and the men to aid in erecting the stamps and to defray the expenses of Forbes and a helper from Belleville to the mine and return.[101] Williams refused to pay for the batteries without a thorough testing, however, and he insisted that the stamps should first be tested at the mine for one month free of charge to demonstrate the validity of the inventor's claims.[102] The machinery had to process at least twenty tons running for twenty-four hours, the stamps "not to get out of order or wear away on an average more than those of an old fashioned stamp mill."[103] In addition, Williams requested that the pattern of the shoe and die should be furnished along with the stamps so that these parts could be replaced locally. Macdonell and Foster agreed to all of this. In reporting to the inventor, Foster cautioned him to have the mill complete in every way so that chances for successful results would not be reduced. Foster's explanation of Williams' situation is exceptionally insightful, with broad implications for mining operators in this district that would be difficult not to see.

> [Williams] has established himself with the old sort of machinery, and by experience gained under considerable difficulties for want of capital has arrived at the point of developing to a paying condition and he fears the risk of a new machine that might make him lose money, which would be ruinous to him in his circumstances.[104]

The stamps took a month to arrive at the mine, and the trial began in February. They proved a dismal failure, so the order was cancelled. There was too much elasticity in the stamps, which caused a loss of power, and the battery needed too much steam, hence fuel.[105] In desperation, the patent agents suggested modifications to the mortar and the steam connection to overcome these difficulties, but the inventor was more interested in promoting his stamps in the larger United States market along with his new grinder and amalgamator that were to be used in connection with the stamps.[106]

When the experiment with the new crusher failed, the old stamps were returned to use, along with three more batteries of five stamps each purchased from a neighbouring mine.[107] Next, Williams replaced the primitive open-kiln roasting with a revolving cylinder furnace custom-designed and built for the mine by its manager, J.H. Dunstan, and patented in 1872.[108] It is interesting to note that Dunstan had spent the previous season overseeing Nightengale's quartz mill.[109] While Dunstan's background in metallurgy is unclear, it must have been considerable for the Dunstan Furnace combined the principles of such previous successful American furnaces as the Stetefeldt, Keith, Wheppley and Storer, and Brückner, together with small improvements of Dunstan's own design. Its power requirement was a remarkably low 2½-horsepower, so that the same engine that worked the stamp mill could work the revolving cylinder furnace, but like the Forbes quartz crusher the Dunstan Furnace consumed too much fuel – one cord of wood every twenty-four hours.

Apart from the Dunstan Furnace, the concentrating and amalgamating operations were fairly standard. Following roasting in the furnace, the crushed ore was mixed with water and the mixture sent over a set of blanket troughs where the gold-bearing particles were drained and separated. Next, the concentrates were treated in amalgamating pans of the Wheeler pattern, popular in the American West. The arsenic in the ores had the unwanted effect of weakening the mercury on the amalgamation plates.[110] Pan amalgamation was everywhere proving of little use in treating refractory ores; it caused losses of up to 75 per cent of the assay values. Other interesting processes were attempted, such as chlorination, and E.J. Chapman's, which included producing a paint during the treatment.[111] But economical and efficient methods for treating the difficult gold quartz continued to elude researchers and methods of concentration and amalgamation remained the chief difficulty.

Gatling/Canada Consolidated Mine

The Dean and Williams operation had taken the lead role in introducing new and experimental technology to the district. Because it was largely unsuccessful, it had the effect of discouraging experimentation and development at other gold mines. By 1871, work at many operations in the district had ceased altogether.[112] At the nearby Powell Mine only surface ore was worked. The cautious owners had installed a small, portable stamp battery and 20-horsepower engine for testing and a small, "ordinary" reverberatory furnace for roasting their ore.[113] The Feigle and Gladstone mines presented a similar picture. At a third neighbouring mine called the Gillen, the superintendent was even less anxious to undertake financial risks prematurely. Like the Dean and Williams

Mine, the Gillen employed a battery of recycled stamps.[114] The entire operation was powered by water, rather than steam. When the dam went out in 1870, work at the Gillen ceased.

Tests in the Dunstan Furnace at the Dean and Williams Mine in 1871 showed that the ore from the Gillen was quite lean.[115] A more scientific assay by Professor E.J. Chapman, however, indicated the ore's general richness.[116] Rather than develop the property themselves, the owners took swift advantage of the encouraging news to unload the property on W.J. Gatling and partners, who already had begun developing a neighbouring gold location and village, Deloro. One thousand tons of ore were mined and stored, and a twenty-head stamp mill was installed, perhaps in the hopes of attracting new investors. Both Gatling properties contained ore which assayed at $14.00 to the ton, a level at which moderate profits were possible. But the failure of the Dean and Williams operation to solve the economic and technical difficulties of separating the gold from the sulphide of arsenic, coupled with the economic depression, soon ended work at the Gatling operations as well. That was not the end of Gatling and his partners, however. Like Gillen before them, they formed a company and attempted to unload their mining properties.[117]

One final attempt was made to revive the gold operations in the Madoc-Marmora area when the Gatling property was finally sold in the economic resurgence of 1880 to a group of New York capitalists, who reorganized it as the Canada Consolidated Gold Mining Company.[118] By then, the mine had been idle for almost a full decade, and so its rejuvenation would be expensive. However, the property now enjoyed a new asset – the Central Ontario Railway line just ten miles distant. Preliminary assay tests also showed the site to be fairly rich with gold. Canada Consolidated hired a remarkably large force of one hundred men to renovate and improve the property, to erect numerous mine buildings and workers' housing, and to outfit it with the latest milling equipment, which included a large chlorination works. Several new shafts were sunk, one reaching two hundred feet, and steam hoisting equipment was installed. Good details about the metallurgical practices at this operation have survived.[119] Ore from the mine travelled an inclined tramway fifty feet to a Blake Crusher, and then the broken ore fell to a series of crushing rolls. Next, the ore received a third-stage grinding in a second set of steel crushing rolls. The material was hoisted back to the top of the mill and passed through a series of sizing jigs, where it was sorted according to shape and density. The coarse sizes were concentrated more finely in Hartz jigs, which were fine-grind jigs developed only a few years earlier at a St. Louis, Missouri, lead works.[120] Next, the sized ore was roasted in a single, lead-lined roasting cylinder. Finally, the roasted concentrates were revolved with free chlorine gas under pressure in a barrel, then washed, and the gold and arsenic were separated from each other – this following the conventional California practice of chlorination.[121] Although this plant, whose capacity was five tons per hour, was much more advanced than any other in the district, it was a commonplace operation by the mining standards of the American West.

To make this operation pay, the owners tried to recover the arsenic by-product that would otherwise have been wasted. Production of arsenic was stepped up in 1884 and 1885 by replacing the old roasting and chlorination furnaces with ones of greater capacity, and a new refining furnace with chambers for the recovery of the arsenic also was installed.[122] This was the final development at the location, however. Assays performed in Boston and New York indicated that while as much as 95 per cent of the gold could theoretically be recovered from the concentrates, there still was so much difficulty in treating the ores – particularly separating the gold from the sulphide of arsenic – that the chlorination process was a failure. The new works had succeeded in raising the value of the concentrates from $14.00 to $20.00 per ton, yet this increase did not begin to offset the increased costs – especially of chlorine – to produce it.

The Canada Consolidated operation managed to limp along for several more years since it was the only serious producer of arsenic in the province. Arsenic was still being derived from the mine refuse and mill tailings in 1888 when the operation was visited by members of the provincial Royal Commission on Mineral Resources, though the mine itself had been idle and full of water since 1884. The new recovery process being used was developed by a Madoc engineer, Charles Taylor, who had gained his experience in treating arsenical gold quartz in the goldfields of Nova Scotia. The tailings were run over an oscillating table containing mercury, sodium amalgam, nitrate of mercury, and copper filings; from this process the company obtained an average of $4.50 per ton from the refuse tailings, $15.00 to the ton of ore, and $60.00 to the ton of concentrates from this and other mining operations in the area. Yet, too much gold was still being left behind in the waste material. Incidentally, early in the 1890s the Hastings Mining and Reduction Company, a group of Toronto and Philadelphia businessmen, organized themselves for the purpose of testing an experimental, more sophisticated chlorination process here. It was the Walker-Carter process invented by an American named Bancroft, but, like the earlier ones, it proved not to be the hoped-for breakthrough.[123]

By now the southeastern district's limitations for mining had become well-known, both from experience and, in the case of iron ore, through explorations by diamond drill. Developers were only too aware that the application of capital and technology were insufficient to overcome the problems facing mining in the district. The principal problem was the poor prices for the minerals of this district on the market, and it was not easily overcome with applications of technology. Surface showings of great initial promise yielded small deposits that were irregular even at shallow depths, and in general ores were low-grade and extremely difficult to treat using the technology of the day. Only those operations that could be conducted on a small, simple scale and that enjoyed cheap transportation costs – seldom did that mean rail transportation – and assured markets could be considered worthy of continued operation.

6

WESTERN PENINSULA DISTRICT

Beginning in 1859, the Ontario oilfields, together with those located in the conterminous territory of Pennsylvania, were the first in the world to be exploited on a commercial scale. The experience with mining and technological change in this district differed markedly from that of the other two districts. Here minerals were extracted in fluid form as liquids by means of wells, rather than by mining or quarrying stationary, mineral-bearing rock from the earth. Whereas the uses for and methods of securing the other minerals were well known, petroleum was a new and relatively unfamiliar commercial commodity. The novelty as well as the peculiarities of petroleum, such as its flammability, plus the practical need to process the petroleum and salt fully because their markets were largely domestic, led to a great deal of experimentation and innovation in this locale.

The chief uses of petroleum were as an illuminant and, secondarily, as a lubricant, though some attempts were also made to use it locally as a cheap source of energy fuel. While the Ontario oilfields were not especially productive, as a consequence of the experimentation and innovational activity, Ontario had an opportunity in the field of petroleum technology to make important contributions of special mining technology and skilled labour at an international level. Moreover, drillers who explored deeper and further afield for additional petroleum deposits soon discovered what was, and remains today, Canada's major commercial salt field. Salt production in the form of brine was technologically close to the petroleum industry; some of the technologies developed for petroleum production were successful when applied to salt, and the salt producers and manufacturers themselves were very inventive when it came to dealing with processes for the recovery of salt.

The petroleum and salt district lies at the western end of the St. Lawrence

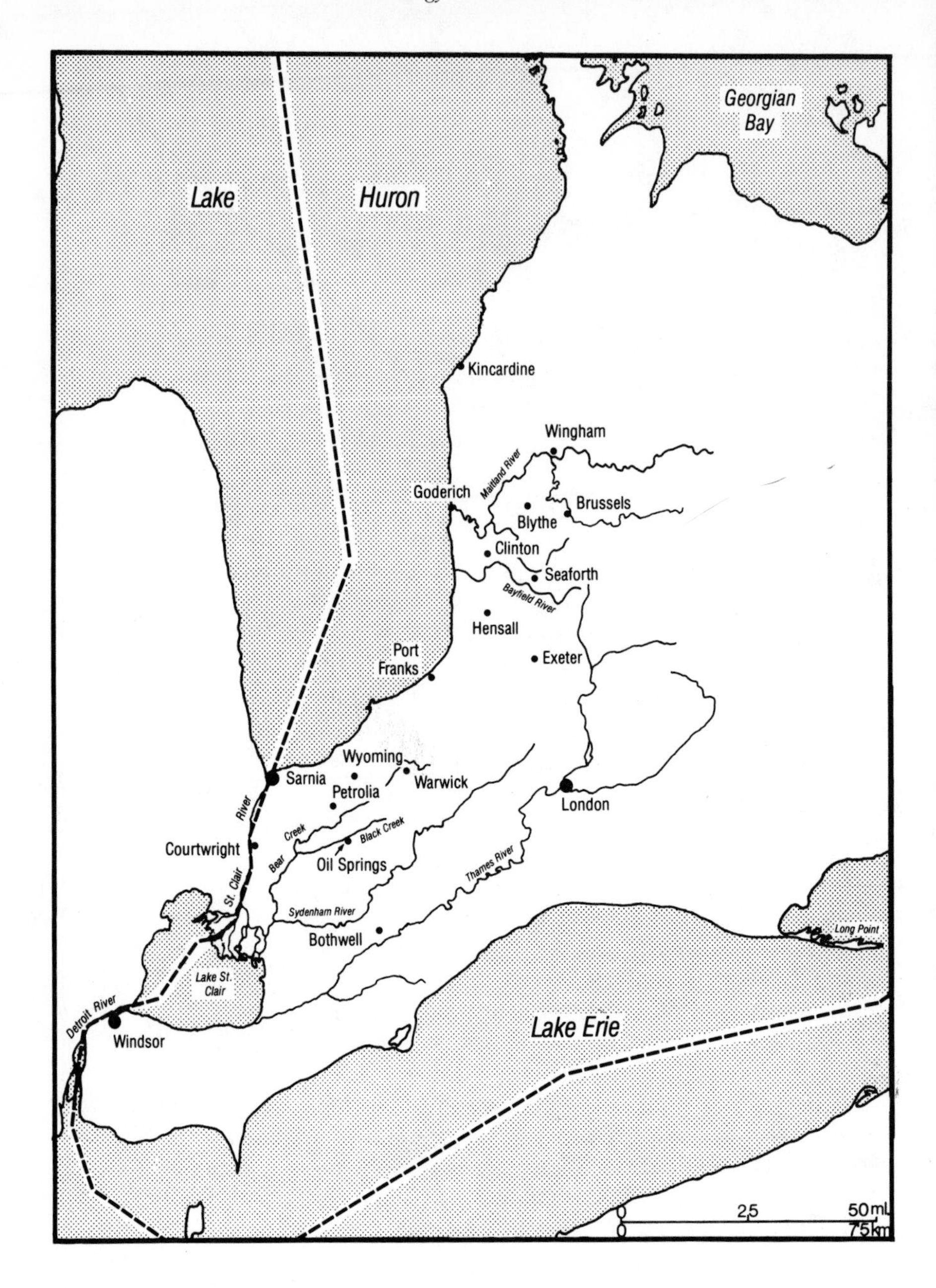

MAP 5. The Western Peninsula Mining District.

Lowlands in a section which geologically is comprised of comparatively undisturbed limestones and Palaeozoic strata rich in fossils, overlaid by drift clays, sands, and more recent surface deposits. It has a remarkably even surface, except where cut by rivers, is covered with prime agricultural land, and is only thinly forested. The westerly region near the shores of Lake Huron contained most of the commercially valuable deposits of petroleum and salt. The petroleum occurred in three horizons in two distinct areas in Lambton County. The first area discovered, at Oil Springs, extended over a small territory two and a half miles by one mile and at less than four hundred feet. The second, more extensive one, at Petrolia, stretched thirteen miles by two miles and was slightly deeper. Petroleum also was found in other localities over a much wider area, but not in paying quantities. The salt area occupied the northeastern edge of beds running south and west into Ohio and Michigan and lay much deeper than the oil-bearing formations. The primordial seas that had once covered the district had deposited great beds of salt in two horizons in the counties bordering on Lakes Huron and St. Clair. The beds were thickest in the Sarnia and Goderich regions – ranging from 140 to 600 feet in depth – but they occurred only at the base of the formation, beginning at a depth of roughly 1,000 feet.

The character and mode of occurrence of both minerals presented serious difficulties to operators wishing to exploit them. The petroleum fields were quite small in extent, and even more serious, unlike the case of Pennsylvania crude, the high sulphur content of the oil defied treatment. This circumstance rendered the district's petroleum somewhat impractical as an illuminant – its chief commercial use. On the other hand, the petroleum deposits were relatively easily and inexpensively reached: they were quite shallow-lying, and they occurred in nearly horizontal strata so constant in thickness that drillers could count on easily finding the same beds at approximately the same depths elsewhere in the field. When the pioneer oilfields declined by 1865, a spectacular strike in 1866 at Petrolia, a little to the northwest, touched off a second boom which resulted in a longer-lived oilfield consisting of several thousand wells. In the long run, however, the Ontario petroleum wells were low-yielding and quickly had to be supplemented by large numbers of new wells.

With salt production the situation was slightly different. The salt resource, while virtually limitless in extent and of high quality, occurred at more than twice the depth of petroleum. Sinking brine wells therefore was more costly than sinking oil wells. Moreover, salt occurred in association with gypsum, a mineral that appeared in the brine and tended to reduce its purity and the commercial value of the recovered salt. The gypsum also encrusted the pipes and evaporating pans so heavily that encrustations had to be regularly removed – which affected production and increased the maintenance costs. Since brine wells seldom ran out, no more than a dozen brine wells in all were operated in the district at any given time.

Plate 13. "Shooting Off" and Oil Well Using Nitroglycerin, Oil Springs, c. 1870s. *The oilfields of Ontario and Pennsylvania were the first in the world to be exploited on a commercial scale. In the long run, the Ontario petroleum wells were low-yielding. It became the practice to rejuvenate old wells and activate new ones with torpedoes employing dynamite or nitroglycerin to fracture the rock underground. This method, which allowed the petroleum to flow freely, was imported from the United States.*

The setting for mining in this district was considerably better than in the other regions. The mining occurred in a relatively well-populated farming and emerging manufacturing region which did not present the same physical, social, and economic difficulties to operators as those in the province's other two districts, located in Canadian Shield. Towns and villages already existed throughout the area. The oilfields were only about fifty miles from the St. Clair River and within twenty-five miles of Lake Huron, and the salt district was even closer to the lake.

The western peninsula was well served by roads and even by railways by 1861, the time of the first oil boom. Unfortunately, these main lines did not run through the oilfields themselves, so during the early boom years crude oil was carted with difficulty through swamp and forest over a thirteen-mile plank road built for the purpose from the Oil Springs field to the nearest railway station, Wyoming. Beginning in the mid-1860s, the four railway lines already traversing the country made plans to build branch lines to the area, largely in response to the petroleum boom. The Great Western beat its competitors in reaching Oil Springs by running a thirteen-mile branch line from its Sarnia Branch at Wyoming Station in 1865. As soon as serious deposits of oil were discovered at nearby Petrolia, businessmen there financed a five-mile branch line to the Wyoming Station.

Petroleum-refining could be done closer to the markets and at convenient collecting centres. The Ontario refinery business expanded enormously over the next decade and reached its peak with fifty-three refineries by 1870. Then, newly opening oilfields elsewhere in the world, plus the formation of the Standard Oil Trust in the United States, effectively eliminated export markets for Ontario petroleum. Refineries were no longer being located at Oil Springs but, rather, along the rail lines in towns and cities throughout the district and in principal manufacturing and distribution centres, such as Hamilton, Toronto, and Montreal. The greatest number of refineries, however, until 1878 were situated in nearby London, the regional centre. After 1878, when the Petrolia field had come into its own and the Michigan Central Railway (Canada Southern) had run a branch rail line through it, Petrolia replaced London as the regional refining centre. Salt, on the other hand, was so bulky and of such low value in relation to weight that it usually did not even pay to ship it by rail. Consequently, the most profitable salt works, which were always located near the brine wells, were at or near the lake port of Goderich.

THE PIONEER PETROLEUM ERA, 1850-1865

In the first years of exploration and exploitation the petroleum industry was a wide-open, unorganized, highly experimental enterprise largely undertaken by inexperienced local entrepreneurs. Oilfields being opened up at the same time in Pennsylvania, then in West Virginia, Indiana, Kentucky, and Ohio, were similarly unorganized. The pioneer phase everywhere was characterized by peaks of great discoveries followed by periods of low production.

The first oil discovered by digging a well occurred in 1857 or 1858, in what were referred to as mineral pitch or "gumbo beds" found earlier and used for medicinal purposes by early settlers. The first official notice given them in Ontario was by Alexander Murray and T. Sterry Hunt in Geological Survey reports for 1849-50 and 1850-51.[1] Analyses by Hunt, the Survey chemist, pointed to the suitability of these deposits for use as asphalt. Murray was a resident of Woodstock, so historians believe he must have passed the information about the gumbo beds and their commercial potential to another resident of that town, George Tripp, who subsequently made the first attempt at commercial exploitation of the oil springs of Enniskillen Township.[2] Tripp organized an International Mining and Manufacturing Company to explore and work the asphalt and other minerals on properties it had acquired in the Townships of Enniskillen, Dawn, and Brook in Lambton County.[3] The company, backed by Woodstock, Hamilton, and New York City financial interests, attempted to refine the oil gumbo dug from the beds in the manner of extracting peat or bog iron ore. The venture had failed altogether by 1857, apparently from the lack of proper refining technology, inadequate transportation, and insufficient capital.[4]

While Tripp was attempting to exploit the deposits of oil gumbo, a Nova Scotian doctor, Abraham Gesner, was in the United States publicizing his revolutionary new patented illuminant "kerosene," which represented the first commercial use of petroleum and thus was an incentive for developing the wealth of petroleum resources because they could be produced from many different hydrocarbons.[5] When James Miller Williams, a highly successful entrepreneur from Hamilton, acquired the Tripp company's assets in 1859, he proceeded with Tripp's help to dig wells for free-flowing oil, rather than simply attempting to refine the impure surface gumbo as Tripp had tried to do.[6]

In the Enniskillen area, on the shores of Bear Creek, oil was struck by shallow wells dug to a depth of forty to sixty feet in the clay. On the south bank of the creek, Williams built a crude refinery, basically a retort in which he distilled oil to produce kerosene.[7] Other wells followed. These surface wells were four to five feet in diameter with sides cribbed to prevent them from collapsing, and they were driven down to the bed of gravel that covered the surface rock where the oil would press into the well.[8] In short order, Williams sunk artesian wells into the rock layers which constituted the middle horizon of oil and achieved good results. The practice of digging surface wells was not entirely displaced by artesian wells sunk to lower levels, since the former were so much cheaper to construct by farmers and small-scale producers who could hope to make some living from them. Such surface wells were still being sunk as late as 1866.[9] As word of the petroleum discovery spread, new adventurers attempted to enter the field, only to discover they had to purchase small plots at high prices from Williams and his associates.[10]

The activities in the Ontario oilfields in the first years simply paralleled developments in Pennsylvania and other North American areas opening to development. The success of the Enniskillen field and the monopoly on the oil lands led outsiders to permeate the deeper-lying rock strata, which in 1860 and 1861 revealed a great flowing region.

Hugh Shaw's celebrated gusher was struck at the 208-foot level in 1862 at what became known as Oil Springs.[11] This led to the first real oil boom and to serious speculation in oil lands in Enniskillen Township. This well and others in the six-and-one-half-acre field flowed spontaneously, but the flow – which arose from discharges of pockets of natural gas imprisoned in the oil pools – quickly diminished or ceased altogether, whereupon the oil had to be pumped to the surface. The initial potential was an impressive two to five thousand barrels per day.[12] By this time Williams had shifted his refining operations to Hamilton, where his Canadian Oil Company, with "16 hands," produced 120 barrels of oil per week.[13] Soon after these important discoveries of petroleum were made in the underlying rock at Oil Springs, wells were sunk across three acres at nearby Bothwell Village and at Petrolia to the north where surface indications had also been observed. The experience was similar to that of Oil Springs. Not until the discovery at Petrolia of the King Well, in 1866, however, did the presence of oil in the third, most extensive strata become known.

Most of the technology appropriate to drilling, pumping, storing, and transporting the crude oil and refining the petroleum were worked out in Ontario and Pennsylvania during this early period of exploration, experimentation, and initial development. But no serious development could take place until drillers and producers had a better understanding of the nature of petroleum and of the mode of its occurrence in this district. This was where geologists and the Geological Survey of Canada were indispensible. By lowering information costs, the Geological Survey reduced the financial risks for petroleum operators during the high-cost phase of exploration and initial development.

Numerous local theories about the extent and location of the productive territory had been speculated upon during the first few years, and courses of action were based on them. Conventional wisdom dictated that the petroleum originated in the oil-soaked strata in which it was found. Indeed, that belief was not easily dispelled, as the testimony of prominent pioneer oilman, John Henry Fairbank, to the Ontario Commission on Mineral Resources in 1888 reveals: "I think the oil is on [sic] a certain saturated area, and not that it comes from below at all."[14] However, the rise of the Ontario and Pennsylvania oil industry had led T. S. Hunt and others in 1861 to study the mode of occurrence of petroleum and its relationship with adjacent geological structures.[15] Hunt was correct in his belief that petroleum should be sought at the top of a major anticline capped by an impervious stratum, that it originated from rocks below, and that it merely permeated the strata through which it migrated, rather than originating in them. His work on petroleum eventually gained international acclaim.

In the first few years, oil operators would crib down to the bedrock, which lay at a depth of between forty and one hundred feet, then drill down another three to four hundred feet to reach oil. Using the primitive spring pole method, drilling such wells took upwards of six months and cost several thousands of dollars, without any guarantee of a return on the investment.[16] Once the spring pole method was superseded in a few years by steamdriven drilling rigs operating on the spring pole method but using heavy cable tools of the sort developed and employed in the Pennsylvania oilfields, the time required to sink wells was reduced to a period of a few months and the cost of sinking to $1,500 on the average.[17] Moreover, the business of drilling wells fast became a separate, highly competitive business requiring considerable skill on the part of the drillers. In all, about a thousand wells were drilled in the Oil Springs field during its five-year boom phase.

During drilling, a well had to be cased and packed. The well-packing in this district usually consisted of leather bags filled with flax seed, such as Hugh Shaw had used to pack his celebrated gusher.[18] When wet, these would swell and fill the cavity. They were not permanent, but then neither were these wells in most cases. Thus, while occasional experiments with more sophisticated, permanent well-packers continued to be carried on, seed bags remained in

Plate 14. "Kicking Down" an Oil Well Using the Spring Pole Method, Oil Springs, 1860s. *Depicted in this photograph are some of the pioneers of the petroleum industry of Ontario. The first oil wells were dug as surface wells, then artesian wells were sunk to deeper levels by being kicked down by manpower using the primitive method of spring poles.*

general use in the Ontario oilfields at least until 1890.[19]

Pumping the crude oil to the surface presented little need for new technology; regular artesian well equipment – valve pumps for shallow wells, lift or force pumps for deeper ones – were used.[20] The shallow wells of the middle horizon could be pumped using the same spring pole method as for sinking wells. About 1863, as the deeper levels were reached, this method was superseded by steampowered pumping units. At first a separate steam engine supplied the power at each well, but as the number of wells increased and production began to fall off between 1863 and 1865, their relative scarcity, along with the high cost in terms of fuel and labour of operating them, inspired a drive for a more practical power system.[21]

As was the case with underground mining, the solution lay in a central power source with a practical power transmission system. Local newspapers reported some improvisations using a single steam engine to power more than one operation had already taken place; two wells were pumped by a single engine in 1861, for example.[22] In the next year, a single engine was employed to operate a sawmill and pump a well, while another producer used the same engine to pump one well and drill another. This trend resulted in the introduction in the mid-1860s of the jerker system, which as described earlier, made it possible to pump dozens of petroleum wells with a single small-power steam engine (central power house).

The pumping methods on the shallow-lying, low-yield oilfields illustrate the persistence of a relatively primitive technology where there was no clear economic advantage in increasing the rate of flow.[23] John Henry Fairbank, who was possibly the most successful and celebrated entrepreneur/politician of the Ontario oil industry, was the first operator to use this system successfully on a large scale in the western oilfields, about 1863.[24] His biographer points out that the enterprising Fairbank seems not to have considered this important enough to note in his diary, which otherwise is very thorough.[25] In his testimony to the Royal Commission on Mineral Resources, Fairbank credited a Mr. Reynolds with having introduced one critical component – the horizontal wheel – to the district.[26] Fairbank and his contemporaries reported that the entire jerker system was unique, even a local invention, but, as already suggested, it was a widely used principle in seventeenth-century Europe applied to transmit power from water wheels to pumping water from deep mines.[27] Although it is not known how it was that the Ontario (or Pennsylvania) oilmen learned of this method – the *Stangenkunst* – or precisely where it was introduced first in North America, the principle would have been a simple one to imitate.[28]

Plate 15. Surviving Examples of the Jerker System for Pumping Petroleum Wells in Oil Springs, Ontario, 1979. *In the marginal oilfields of the western peninsula of Ontario, the old jerker system with wooden rods and iron wheels persisted – even to this day. The horizontal "spider" wheel shown in this photograph by Ralph Greenhill swings through an arc to give the jerker rods the necessary forward and backward motion to produce a stroke at the pumps at each well. The wheel also serves to subdivide and redirect the power being transmitted to the wells from the central power plant. The wheel was said to have been introduced to the oilfields by a Mr. Reynolds.*

Once the oil was pumped to the surface, it had to be collected and forwarded to a refinery for processing. Each well had a small-capacity tank for the temporary storage of the pumped oil, whence it was collected by gathering lines or teamsters and stored centrally in larger containers. A great deal of wastage occurred in the early years because of inadequate controls over the flowing wells and the inability to store the crude oil properly. The first tanks used were small wooden ones that rested on the ground. The crude petroleum was refined at Oil Springs or shipped in barrels over the plank road to Wyoming station, a dozen miles distant; from there it was moved by rail to refineries located at London and Hamilton or was shipped to more distant markets. Wooden barrels were the most commonly used containers, but they posed difficulties because the oil leaked from them – which was of particular concern to the rail operators who shipped them. Also oil permanently spoiled barrels for any future use as containers. Used barrels containing other foreign matter were equally unsuitable because they could contaminate the oil.[29] Numerous barrel-sealing compounds such as glues and isinglass were experimented with in Canada and the United States. Otto Rotton, a Kingston physician, obtained six patents in connection with his "Union Cementing Process," for a process, materials, and apparatus with which barrels could be made impervious to petroleum and similar products.[30] His patent was quite similar in principle to the Woodruffe patent, already awarded in the United States to a resident of Albany, New York, and publicized a few months earlier in the *Scientific American.*[31] It is not known whether Rotton's patented process was ever tried or not. Certainly the problem of preparing wooden barrels to hold oil was never really solved.[32]

During the first years production of illuminating and, to a lesser extent, lubricating oils dominated the petroleum-refining business; the other petroleum products were marketed as they were secured, without further processing, or even dumped as waste.[33] The only exception to this emphasis on the production of illuminating and lubrication oils was the numerous experiments with refining to manufacture illumination gas and paint from petroleum.[34]

Illuminating oils were the most profitable petroleum products, but Ontario oils burnt with an offensive odour, encrusted the lamp wick, and produced smoke which blackened the glass chimney. Hence, modifications in distilling and refining the crude, and especially the particular brand of petroleum found in the western peninsula, were quickly made. A great deal of secrecy surrounded this experimentation, but the patent record indicates the general trends. Several improvements in petroleum stills were patented by local residents in these first few years, though it is not certain whether any of them apart from one patented by the oilman and merchant Hugh Shaw were actually employed there.[35] Shaw's Improved Dome Petroleum Separator was a double still in which were placed a fine wire screen and a coarser brass mesh plate through which the petroleum vapour had to pass before reaching the worm. This feature was intended to screen out the impurities. The inventor claimed his apparatus was an improvement over the Hetfield Dome then used in the United States.[36] John Fleming, an engineer from the village of Petrolia, patented a "Double-Acting Petroleum Still," while Charles H. Waterous, a Brantford machinist and founder of a major iron foundry, patented the "Waterous Centripetal Churn and Centripetal Agitator for Refining and Fitting for use Rock Oil."[37] Most of the patents for improvements in the refining of petroleum were aimed at improving the burning quality – especially the odour – of illuminating oils, the majority being proposed by chemists and refiners.[38] Finding a workable method for deodorizing petroleum would engage innovators for several more decades. In all cases the refining required large quantities of sulphuric acid, so it only made sense for refiners to try to recover that commodity for reuse. The first such process was patented by the Oil Springs refiner Robert Loftus in 1864.[39]

As a final note on innovations in refining during the period prior to 1868, when government regulation and inspection of the refined petroleum came into being, refiners were free to manufacture the crude to any gravity oil they desired. The lighter oils were the easiest to produce, but they thickened too easily in cold weather owing to the presence of paraffin, and they also ignited at dangerously low temperatures. In all, without quality controls the oils produced during the pioneer phase of development were rather unsafe to use.[40]

THE PETROLEUM AND SALT ERA, 1866-1890

The Petrolia Years

Following the discovery of flowing wells at the lower horizon at Petrolia in 1866, the Oil Springs fields were all but abandoned.[41] Better business organization and somewhat improved technology made petroleum production and refining more efficient and less costly. Better techniques for sinking and pumping wells (which took on the characteristics of a distinctively Canadian system), for storage of surpluses, transportation by means of special conveyances, and refining were all developed and propagated. New uses and surer domestic markets for Ontario petroleum products were found, and the dominion government involved itself in protecting the domestic markets for the industry, regulating petroleum-refining, and for a short time in response to the new national policy of protective tariffs in the early 1880s, even sought out new domestic markets for Ontario illuminating oils.[42]

In the long run petroleum wells were low-yielding and quickly had to be supplemented by sinking new wells – by 1880 as many as two hundred wells were sunk annually – discovering new ways to reduce the costs and simplify the well-boring and pumping processes, and blasting out "dry holes." Basically, the techniques of oil-well drilling remained as before except that the surface material was now drilled and cased, rather than dug and cribbed, and, as described in detail earlier, a pole-tool method of drilling through rock superseded the use of the cable tools that remained more popular in the deep-lying Pennsylvania oilfields. The pole tools were both efficient and inexpensive, and the poles themselves lasted through two or three years of constant use, although they were unscrewed and screwed together again very frequently.[43] Well-drilling became a highly specialized business, and as drillers became more expert and the equipment more efficient, the time and expense of sinking a well was greatly reduced from months to four or five days and the cost to only a few hundred dollars.[44] Eyewitnesses of Enniskillen drilling operations in 1872 reported that a single producing hole in ten would be enough to repay a well-owner.[45]

The man who introduced this pole rod technique, William MacGarvey, had moved as a boy with his family to the village of Wyoming, in 1857, and he entered business there in 1861. Following the rich Petrolia strike in 1866, he moved to that town to become involved with oil, and he introduced his pole-tool technique in the same year. A modified version of the Fauvelle system of drilling with hollow rods had been experimented with at Oil Springs the previous year.[46] It is not known whether MacGarvey was involved with this activity, but it was reported locally in the newspaper, so he was likely aware of it. Like John Fairbank, MacGarvey became a moving force in the oil business. He improved the technique and he built up his fortune over the ensuing decade to the point where he took his pole-tool percussion rigs and a gang of drillers to Europe to work on the new oilfields opening up there. By 1880, he was a principal owner of Galician oil wells. With further improvements made to suit European conditions, the Canadian system became the main well-drilling rig employed in Central Europe, and MacGarvey became an extremely wealthy and powerful man.[47] The Canadian drilling rigs developed and manufactured in southwestern Ontario were well-suited to small scale, pioneering stages of oilfield development in relatively primitive areas. Gangs of Ontario drillers travelled throughout Canada, the United States, Eastern Europe, and Asia; by the 1880s, many also were employed in Australia and Mexico.[48]

As old wells ran out, they were abandoned, and new wells were constantly sunk. As the number and depth of wells increased, numerous patents were granted for improvements to well-boring and pumping techniques. They are particularly concentrated in 1866, when the importance of the Petrolia oilfield (and the Goderich salt region) were uncovered. For the most part these improvements involved tube wells and pump valves.[49] Sometimes the proliferation of new wells interfered with the flows of the existing wells. There were other problems too. The abandoned wells filled with water when the tubing was removed, and the water from them then would appear, unwanted, in producing wells in the vicinity. In fact, the problem created by abandoned wells was so

Plate 16. The Petrolia Oilfields, c. 1870s. *This field came into its own in the late 1860s with the discovery of flowing wells at the lower horizon. With uncontrolled and increasing production and the fluctuating seasonal demand for kerosene, which was the chief product of refineries, safe and adequate storage became a serious concern. A blaze broke out in the Petrolia oilfields once that covered twelve acres and lasted two weeks.*

serious that the provincial government authorized injured parties to fill in or otherwise stop the flow of water in them.[50]

New techniques were also employed for rejuvenating old wells. Mostly these consisted of torpedoes employing dynamite or nitroglycerine to fracture the rock underground to allow petroleum to flow, a method mainly imported from the United States.[51] According to the 1889 testimony of John Fairbank before the Ontario Commission on Mineral Resources, it had become standard procedure by that time to "shoot off" even new wells with nitroglycerine.[52] A number of well rejuvenation devices were patented in Ontario; all but one involved introducing water or steam into wells, which would have been undesirable in the view of other well-owners.[53] The device used at at least one well made it possible to collect the gas escaping from wells and use it both to fuel the boiler for the pump and to assist in the flow of oil in the well.[54] The apparatus was employed at the "Maggie Well" and appears to be similar in principle to that patented in 1866 by Alexander Gale Alexander, a Hamilton plumber and pipe fitter.[55]

This uncontrolled and increasing production, as well as the fluctuating seasonal demand for kerosene and hence for crude oil, made storage a serious concern. The wooden storage tanks littered the fields, constituting a sizable expense (about $5,000 each) and posing a serious fire hazard. One major oil fire in 1866 finally inspired a solution to the problem. It is credited to John D. Noble, an early oil exporter who, like MacGarvey, arrived at Petrolia in 1866. Noble had purchased land, sunk a few wells, and built numerous standing wooden tanks in which to store the crude oil, only to see the entire enterprise go up in flames as the result of a blaze started on an adjoining territory.[56] The fire covered a twelve-acre area and lasted a terrifying two weeks. As a consequence of this experience, Noble devised a system of subterranean tanks that quickly was adopted throughout the district. The idea of using underground tanks was patented in Canada that same year by an American farmer, T.M. Ottley.[57] In applying the principle of underground tanking to the western peninsula district, however, Noble showed ingenuity. The tanks were excavated in the impervious drift clay of the oil belt and lined with wood, making them capable of holding great quantities of oil — crude or refined — at little expense under relatively fireproof conditions. These subterranean tanks averaged thirty feet in diameter and thirty-five feet in depth, although some were as large as forty feet wide and sixty feet deep.[58] An average tank held eight thousand barrels and cost $2,000 to build.[59] Storage outside the clay region, however, still presented a major difficulty. At refineries in London in 1872, the holding tanks were set only partly into the ground, then banked over with earth for safety.[60]

Like well-drilling, laying and operating such tanks usually became a separate business, as did storage, so producers would pump the crude from the wells through pipelines to company receiving stations, where the amounts sent would be registered and credit notes would be issued, much in the manner of warehouse receipts and, later, grain elevator receipts. Holders of the credit notes had the option of cashing them at a bank or selling them to refineries, which could call for the oil on demand.[61]

The crude was transported to refineries in wooden barrels or, as was universal after 1870, metal tank cars via rail lines. This handling plus the railway shipment costs proved expensive to refiners who already were facing a competitive market for their products and thus were attempting to reduce production costs. In line with what was happening in the American petroleum fields to break the control of railway shippers, several pipelines were incorporated in the western peninsula, beginning in 1873.[62] These were to be fairly short lines: the first one was built between Petrolia and Sarnia in 1880.[63] Longer lines seem not to have been practical for the Ontario oilfields. The supplies of crude were very uncertain, and the cost of pipe – which had to be imported from the United States – was high.

Greater changes were made, with even further government involvement, in the treatment of crude by refining. Most of the changes were in response to regulations imposed on the refiners by the dominion government in 1868 when it began inspecting and taxing refined petroleum and laying down regulations concerning standards. To undertake this accurately, the Department of Inland Revenue itself had to work out technologies for testing the specific gravity of various petroleum products and for preventing fires; they are described in the department's annual reports to Parliament. Because the government was moving beyond regulating technological innovations to becoming an innovator in its own right, the activity bears closer examination.

Under the laws and regulations imposed by the first excise and inspection of petroleum act, all refiners and any persons operating petroleum stills had to be licensed and submit returns on their business and the quantity of petroleum (crude or refined) received, produced, and treated by them.[64] Refined petroleum was subject to a fire test, and it was unlawful to produce or store refined petroleum whose temperature could not be raised to at least 115 degrees Fahrenheit without producing a vapour that would ignite or explode at that temperature.[65]

TABLE 2
LOCATION OF PETROLEUM REFINERIES IN ONTARIO, 1870-1890

	1870	*1871*	*1872*	*1873*	*1874*	*1875*	*1876*	*1877**	*1887*	*1890*
Cornwall	1	0	0	0	0	0	0	0	0	0
Guelph	2	0	2	1	1	1	0	0	0	0
Hamilton	5	4	4	4	3	1	1	3	1	1
London	20	20	17	15	16	14	6	18	2	2
Paris	5	4	3	2	2	1	1	5	0	0
Petrolia	0	0	0	0	0	0	0	0	9	8
Sarnia	18	17	16	14	17	9	8	11	1	1
Toronto	2	1	2	1	1	2	2	1	0	0
Total	53	46	44	37	40	28	18	38	13	12

Sources: Canada, Department of Inland Revenue, *Report of the Commissioner* (Ottawa, 1870-76); "Mineral and Mining Statistics," (Ottawa, 1887-88 and 1890-91).
*The year the government repealed the excise duty on Canadian petroleum.

The Ontario refinery business expanded enormously in the 1860s, reaching its peak in 1870, when the refining centre had shifted to London from Oil Springs. After 1878, Petrolia replaced London as the regional refining centre.

The burden on refiners was substantially reduced in 1877 when the government eliminated the excise tax on Canadian petroleum because of the increasing market problem facing the Ontario petroleum industry. The inspections, however, continued, though under new regulations that were easier on the refiners.[66] An 1880 amendment to the Petroleum Inspection Act of 1877 reflected the changing complexity of illuminating oils and gases being manufactured in Ontario and was even more explicit in its restrictions of particular products and tests for their uses, thereby further influencing the direction of technological innovation in refining.[67]

Interestingly, a higher flash test was required of imported petroleum sold in Canada for illuminating purposes. Rather than discouraging sales of foreign oil in Canada as intended, however, this regulation may actually have encouraged foreign sales because it resulted in a higher quality of oil being imported.[68] Overall, the 1880 act imposed new responsibilities on refiners and dealers because by it they took over from the inland revenue inspectors the task of determining the quality of refined petroleum. The inspectors were restricted to spot-checking and otherwise determining the accuracy of the grades claimed by the vendors.[69]

The effect of these federal regulations and a federally sponsored technology on oil refining, as well as the Dominion's commitment (1870-75) to use domestic petroleum in dominion lighthouses, was to generate further domestic refining innovations to improve the quality and safety of the products.[70] The Department of Marine and Fisheries even made a special technological contribution by inventing a lamp that would burn Ontario oils to produce a brilliant light without encrusting the wick or smoking the glass and also by designing a new galvanized metal tank for oil storage at lighthouses.[71] But at the same time the federal regulations adversely affected the petroleum industry by reducing the percentage of illuminating oil that refiners could secure from their crude from 60 per cent to 40 per cent and making the high sulphur content of the products even more noticeable.[72] With the already poor reputation of Canadian petroleum in foreign markets, these changes inspired even more local experimentation with deodorizing the illuminating oils, manufacturing illuminating gas and lubricating oils, and developing commercial uses for petroleum by-products such as tar, which formerly had simply been thrown out.[73]

Of course, the most important difficulty in refining the inferior Ontario crude was its smell, which threatened to ruin its principal market. A lead process had been introduced in 1868 as an improvement on the standard refining process which employed sulphuric acid to remove the tarry materials, then caustic soda to reduce the sulphurous nature of the products.[74] In this new process, tentatively credited to an English immigrant, a Mr. Allan, oxide of lead was added along with sulphuric acid.[75] The Ontario illuminating oils remained so malodorous, however, that of the forty-two patents granted between 1861 and 1890 to Ontario residents for improvements in the art of distilling and refining petroleum, no fewer than seventeen were expressly concerned with deodorizing petroleum.[76] Most were ineffective, however, and the 1868 lead process remained in general use until 1886, when the new but similar patented refining processes of Henry Kittridge and Martin J. Woodward, both refiners, gained widespread, if only temporary, acceptance in the district.[77]

The new refining process introduced by Kittridge and Woodward involved additional distillation to remove the sulphuric acid, lead, and sulphur which ordinarily would have been left in the illuminating oil. It is interesting to note that a redistillation technique was being experimented with as early as 1872, when F.A. Fitzgerald of the Union Petroleum Works, London, Ontario, was seeking ways to raise the flash point of the illuminating oils being supplied by him to dominion lighthouses. With the economic crash of 1873, however, he decided the efforts were prohibitively expensive.[78] The higher standard imposed on petroleum products raised the cost of refining; so renewed attempts were made to recover the chemicals used in refining. For example, the London oil refiner John R. Minhinnick (Empire Oil) patented a process in 1881 to recover the surplus chemicals used in deodorizing distilled petroleum.[79] While it is not certain if Minhinnick's process worked, John Fairbank's 1888 testimony before the Ontario Commission makes it clear that at least some refiners were recovering waste sulphuric acid and selling it to Detroit fertilizer companies.[80]

When Herman Frasch developed the first commercially feasible process for refining and deodorizing this inferior oil, he was working as a chemist and business associate with the London-based Imperial Oil Company.[81] After being developed in the 1880s for Ontario petroleum, the Frasch process and apparatus found its most successful application in the United States. In 1888 Frasch was hired and his patents purchased by his former employer, the Standard Oil Company of Ohio, so the foul-smelling crude of the Lima, Ohio, field could be rendered more marketable as an illuminant. The hiring of Frasch thereby furthered Standard's monopolistic ambitions, which, ironically, even included the eventual takeover of Imperial Oil in the late 1890s.[82]

A second approach to lowering the cost of refining the inferior crude of the western peninsula was to develop fuel-efficiency and intensify energy in the refining end of the industry. After 1870 all refiners fueled their operations with mixtures of petroleum tar and steam sprayed into the furnace. During the peak of rivalry among refiners, two new types of apparatus for intensifying energy were patented – one in 1874 by the prominent London machinist and iron founder Elijah E. Leonard; the other in 1875 by David R. Winnett, a London boilermaker.[83] Leonard's "Perpetual Still and Process of Distilling Petroleum" was a major innovation that consisted of several distinct elements. Heat from the furnace would pass directly through the still by means of tubes; as many as twenty stills could be heated with a single furnace which burned tar drawn directly from the petroleum stills themselves. Winnett's patent consisted of an improvement to the arrangement of the heating tubes. He attempted to increase the heating surface inside an oil still by means of a series of steam tubes fixed vertically in cases and having a suitable intervening air space running the entire length of the still.

Despite these attempts after 1870 to improve the prospects of the Ontario petroleum industry through technological innovations, exports to Europe faced stiff competition from low-cost producers with superior products (especially in illuminating oils) in the United States and other countries. In the face of such competition, the entire industry underwent reorganization, becoming much more integrated, monopolistic, and centralized, and introduced large-batch processing methods.[84] The important task of organizing the oil business has been carefully detailed in a master's thesis on John Henry Fairbank.[85] That biography offers insights into the business chaos which existed in the early years and highlights the inability of small, locally based refiners in the first decade of the industry to survive in the face of competition from refiners in major distribution centres operating superior plants and hiring professional chemists to improve their product. Another historical source, the surviving business correspondence of the Toronto oil dealer in the 1870s, John Fisken, illustrates the serious extent to which refiners solicited their trade, suggested technological improvements for increasing the efficiency of the handling of petroleum products, and discussed the way in which price-fixing on the part of producers constrained their dealings.[86]

It was only a matter of time before petroleum refiners decided to break the control exercised by producers' associations and guard against takeovers by powerful outside oil interests such as Standard Oil Trust. Accordingly, several London and Petrolia refiners amalgamated in 1880 to form the Imperial Oil Company Limited, which by the mid-1880s successfully controlled over one-half of the district's oil production. When the Imperial Oil Company's operations in London were destroyed by fire in 1883, it chose to expand its refinery at Petrolia and to move its headquarters there. By this time Petrolia had become the true centre of production, refining, and innovation.[87]

Imperial Oil was a self-sufficient, innovative operation. Besides its numerous wells, Imperial also operated a steam cooperage where special petroleum barrels were made and a factory for the manufacture of a patented tin oil can.[88] It never was able to extend its dominance, however, and its attempts to guard against takeovers by outside oil interests by mergers may only have made a takeover easier.[89] When in the late 1890s the company was swallowed up by Standard Oil Trust, the Petrolia refinery was shut down and a new one erected at the lake port of Sarnia, which became a permanent refinery centre.

Salt Operations

Meanwhile, a sizeable salt industry had sprung up in the shadow of these petroleum activities. Salt was discovered in Goderich in 1865 in the course of searching a wider area for deeper-lying oil deposits.[90] Small pockets of natural gas also were struck, but these attracted little interest until the late 1880s when extensive, deeper-seated reserves were struck elsewhere in the district near the shores of Lake Erie. Other wells with saline contents sufficient to warrant treatment were sunk and salt works erected shortly afterwards – in Goderich, in Clinton, 1867, Kincardine, 1868, several in Seaforth, and at Blyth, in the 1870s.[91] Other good salt deposits were discovered in the 1880s at Brussels, Warwick, Hensall, Wingham, Port Franks, and Courtwright. As with petroleum, the brine was encountered in wells sunk throughout the district, but usually it was too weak to pay for the fuel required for its recovery. Because of the thickness of its salt deposit and its accessibility to cheap, lake transport, Goderich became the salt manufacturing centre.[92]

At Goderich salt was reached at the 1,000-foot level. Over the next decade the next deeper and more extensive strata of salt were reached.[93] In the spring of 1876, Henry Attrill of New York City acquired a large tract of lakeshore land in Goderich and commenced his celebrated 1,517-foot bore to penetrate all six levels of salt beneath.[94] His work took nine months and was accomplished mainly with a diamond drill, technology which had been introduced to provincial mining only a few years earlier. The record of his boring and the analyses of his salt by the Geological Survey were published by T.S. Hunt in Survey reports and in a major North American mining engineering journal.[95] Attrill's well was the deepest in the district – even as late as 1890. His explorations and discovery of the particular nature of salt reserves in the western peninsula were essential to the future of the industry and especially to decisions concerning the judicious application of techniques and machinery.

The depth at which salt was found posed no particular difficulty because artesian well technology was fairly well perfected by this time, and drilling for salt, like drilling for oil, was a specialized business in the district. In the first few years, the costs of putting down a deep brine well were between $3,000 and $5,000, and it took up to seven months. By 1870 the cost had been reduced by half and the time by one-quarter.[96] As a rule, the driller supplied everything – machinery, fuel, casing. The ordinary process of boring for salt and pumping brine in this district was the same as that practised in nearby Michigan, so it likely was simply copied by the Ontario drillers.[97] A hole about four and one-half inches in diameter was bored to the salt measure; then a tube three inches in diameter was inserted for the pump. The pressure of water, usually naturally flowing but occasionally forced, running down the casing of the pump to the salt deposit forced saturated water up the tube to its own level.

From there, the brine was pumped to the surface at an average daily rate of eighteen thousand gallons, a quantity sufficient to produce at least a hundred barrels of salt. One and a half cords of hardwood, at a cost of $2.50 each, were consumed in the pumping, which meant that the cost of fuel consumed in pumping brine was not a significant factor in the overall cost of production. As in the case of oil wells, the fuel costs of pumping brine wells could be reduced if one engine was employed for several tasks. At least one instance of this is recorded for Port Franks, where a local lumberman, Joseph Williams, sank a well and then pumped it using the same boiler that ran his sawmill.[98] Moreover, he fired the boiler with sawdust produced as a waste product in the sawmill.

From the well, the brine was piped to holding tanks at adjacent salt works, where salt was recovered by evaporation, after which it was packaged in bulk or, most often in Ontario, was ground and graded for table or dairy salt. It was this facet of the industry which contrasted with that of petroleum and in which much technological experimentation occurred. Here the particular nature of salt deposits and their physical setting in this fuel-poor district determined the techniques that should be adopted and the form they should follow. Solar evaporation was not used in the district, no doubt because of the unsuitable climate. Instead, two existing systems of artificial evaporation were experimented with in succession.

In the first few years, the Onondaga system was almost universally employed in the western peninsula district, perhaps because of the proximity of the New York State salt district model.[99] With the Onondaga system, a block of cast-iron kettles was arranged in two parallel rows of 26 to 30 kettles each, arranged over a furnace. The capacity of each kettle was 120 to 140 gallons of brine, approximately 104 kettles and fourteen and a half cords of hardwood being required to produce 100 barrels of salt daily.[100] And these early salt blocks were inexpensive, about $3,000 for a standard block of 60 kettles.

It quickly became apparent to most operators in the Ontario salt district that while the Onondaga method might have been suitable for the weak brines of the region in which the Onondaga Company manufactured salt, the type of kettles it used were wasteful and unfit to process the stronger brines of Ontario. The kettles heated too quickly and soon became encrusted with salt. Since the salt crusts were unmarketable, there was a great loss of salt, heat, and time, and it increased the labour costs.[101]

Following the 1867 visit and report of Dr. C.A. Goessmann, a chemist with the Onondaga Salt Company sent to investigate the possibility of that company's expansion into the western peninsula district, the English system of moderate slow evaporation in broad, shallow pans was accepted as better suited to Ontario operations.[102] It would appear that this system already had been adapted to the Saginaw salt district in Michigan, although Goessmann does not discuss it.[103] Consequently, within two years of the beginning of salt operations in the peninsula district in 1866, an improved English system of manufacturing was introduced at the new Stapleton Salt Works, and within another two years, it was employed in all the salt works but one.

The Stapleton Salt Works that commenced operations at Clinton in 1868 was outfitted with an English-type system introduced by the owner, John Ransford.[104] Each pan measured twenty-one feet by forty feet and fifteen feet deep – standard for the time but slightly smaller than those which would be used in the salt district in later years. Three wood fires were located under the front pan; waste heat escaped through flues to heat the second pan.[105] The daily production of the two pans was seventy barrels, and the savings in fuel consumption were dramatic. The Stapleton Works was the district's largest by 1874. Six other works followed suit and adopted the English, or the English improved, system, Seaforth, Prince's, Dominion, Ontario (the last three were all in Goderich), and the Bruce (in Kincardine).[106] While Ransford does not seem to have patented his particular arrangement, which economized on fuel and used waste heat, many variations on the English system were patented at the time, and no one type predominated.[107]

Hence, the common feature of all the operations after 1870 was that only the English system of metal pans or wooden vats, rather than iron kettles, was employed. This is an excellent example of the process of collective innovation and a rare example in the Ontario mining districts of English technology replacing American, rather than the other way around. Another popular version of the English system was that developed by Samuel Platt, a Goderich miller and leading salt manufacturer, who introduced his own patented evaporating pan to his Goderich Salt Company works in 1868.[108] According to his patent, heat was applied to the metal pans by means of steam under pressure in a steam chamber to save on fuel. Platt also patented a system of superdrying the damp salt by heating it in a cylinder revolving over a fire, then passing it between a horizontal pair of stones and dressing it by revolving reels.[109] With the economic downturn that began in 1873, salt manufacturers quickly adopted Platt's energy-efficient processes. Four Goderich works – the Goderich, Maitland, Victoria, and Tecumseh – used Platt's patent evaporators, and the Seaforth Salt Works had adopted Platt's patented system of superdrying and dressing salt.

By 1874 three other works – the Eclipse, Carter Brothers, and Carter and Company (all located in Seaforth) – were using a system termed the "Runciman Method," whose origins are not clear. This latter system followed standard English practice except that fireplaces were located at both ends of the pans, resulting in more uniform heat.[110] Another three works – Marshall's, Ogilvie's (both of Seaforth), and Ogilvie's at Goderich erected during 1874-75 – employed steampiping in direct contact with the brine.[111] This technique of heating with steam coils was likely borrowed from English salt manufacturers as well, and it represents yet another example of the sorts of attempts to intensify energy that

TABLE 3
ONTARIO SALT WORKS, 1887-1888

Location	*Company*	*Owner/Mgr.*	*Date Opened*	*# Men Employed and Daily Wages*	*Fuel*	*Annual Yield (bbls)*
Goderich	Platt's Salt Works	John S. Platt	1867	15/$1.25 average	cordwood	16,816
Goderich	North American Chemical Works	Geo. Rice	–	20	–	21,000
Goderich	International Salt Works	Jos. Kind	–	35	–	50,000
Port Franks	Excelsior Salt Works	Williams & Murray	1883	15	sawdust	11,000
Kincardine	Ontario People's Salt Manufacturing Company	A farmers' Co-operative	c1868	15	slack coal	4,200
Courtwright	Courtwright Salt Company	J.S. Nesbitt, Sarnia	1884	9/$1.25 average	petroleum tar	28,000
Blyth	–	–	1878	13	cordwood	50,000
Wingham	–	–	1886	13	cordwood	–
Seaforth	–	–	1871	13	cordwood	10,000

Sources: PAC, Department of Mines, Mineral Statistics, RG 87, vol. 27, File 116; and *Report of the Ontario Royal Commission on the Mineral Resources of Ontario* (1890), pp. 187-91.

Salt was discovered in Goderich in 1865 in the course of searching a wide area for deep-lying petroleum deposits. Because of the thickness of its deposit and its accessibility to cheap, lake transport, Goderich became the salt manufacturing centre of Canada. Despite the numerous technological innovations developed locally to lower the cost of producting salt – especially to reduce the cost of fuel for evaporating brine – only a handful of works were able to survive the poor price and market situation facing Ontario manufacturers.

were characteristic of nineteenth-century innovations.

Certain of the innovations in energy-intense, fuel-efficient techniques introduced into salt operations were also designed specifically to handle the difficulties posed by the gypsum content, which would coat pans during the process of evaporation, reducing the efficiency of heating arrangements and contaminating the product.[112] Salt operations were forced to shut down every few weeks to deal with the problem of gypsum. The president of the Merchants Salt Works in Seaforth patented an evaporating device designed to heat brine and also remove the gypsum from it in pipes before it reached the evaporating pans.[113] Called "Hayes' Patent System," it was adopted at the Merchants Works when it commenced operations in 1871, and within a few years, three of the district's largest works were using it – the International (Goderich), Rightmeyer's (Kincardine), and Kincardine. The one remaining operation, Hawley's in Goderich, employed the rather expensive method of heating in copper pans, which were adopted to handle the corrosive nature of the brine supplying the works. Richard Hawley, Jr., owner of the works, made some minor improvements to the system of heating and settling which he patented in 1871.[114]

All these experiments and technological borrowings increased the efficiency of the salt evaporators and reduced the cost of manufacturing a barrel of salt from Ontario brines by fifty cents on average. Whereas two cords of hardwood were required to produce fourteen barrels of salt in 1869, only one cord was required to produce the same amount in 1874. This represented quite a potential savings for salt manufacturers, but these gains were cancelled by the rising cost of hardwood, which doubled in price between 1867 and 1874. In fact, some of the contemporary explorations for new brine wells in more northern areas around Wingham were simply intended to locate salt deposits close to forested areas and, hence, to cheaper sources of cordwood.[115] As soon as markets and prices for Ontario salt began to decline in the mid-1870s, however, manufacturers began tackling the problem of finding a substitute for cordwood, which was becoming a scarce as well as costly fuel in most of the district. To this end, Martin P. Hayes, the inventor of the popular system of evaporating described above, experimented with burning local liquid petroleum for fuel in the early 1870s. His report to the Geological Survey indicates that petroleum fuel was no more practical than cordwood, though he cautioned that the experimental nature of the fuel made the comparison unfair.[116] By the 1880s, some works employed slack coal (half bituminous, half anthracite) imported from the United States, but they found it costly because of the import duty.[117] Petroleum tar, a waste product of the nearby Petrolia oilfields, was being burned by the Courtwright Company in 1885, though the operators complained to the Ontario Royal Commission on Mineral Resources that petroleum refiners were taking unfair advantage of this local market by doubling the price, charging them the same price for coal as for what was essentially a waste product.[118]

The most satisfactory of all solutions to the fuel problem for salt manufacturers was found not so much in the form of cheaper fuels, but in using the recycled steam from other productive local enterprises. As soon as heating with steam was introduced into the district salt works in 1868, a number of salt works were combined with saw- or flour-milling operations, using the latent heat in steam condensed in an engine to heat the brine.[119] This proved the most efficient technology of all on the fuel side. The owner of a steam-flouring mill in Seaforth, William Marshall, was the first in his district to use steam from his mill engine in the manufacture of salt.[120] In 1874, Marshall employed a system of wooden vats lined along the bottom with galvanized iron steam pipes. Using wood instead of iron for the vats, which was possible because the heat was applied in the form of steam pipes rather than a direct fire, as well as recycled steam, drastically reduced the capital investment required and the operating costs. Marshall reported to the Geological Survey that the profits from his salt paid the entire cost of fuel (which was coal) for his flour-milling operation. The optimism generated from this new system of evaporating and energy-efficiency led Ogilvie's, the celebrated Montreal flour-milling firm, to erect two large steam flour mills with adjoining salt blocks on the Marshall plan in 1874 at Goderich and at Seaforth.[121]

Despite unlimited supplies of relatively pure salt and strong brines and all the attempts to keep pumping and manufacturing costs down by experimentation with borrowed technology and local inventions, this, the country's pre-eminent salt-bearing area, failed at the time to develop a major salt-producing industry. Technological changes that had been introduced to counterbalance the high and rising costs of salt production had been successful enough. The real problem was one of markets. Salt was more expensive to produce in Ontario than in the United States, added to which were the freight charges and variable duties on Canadian salt, so Ontario salt manufacturers could not look forward to markets in the United States.[122] Even more serious, like the Ontario petroleum producers and refiners, Ontario salt manufacturers had to struggle even to retain a share of the domestic market. In the case of salt, this situation arose because of the competition from cheap English salt shipped over as backhaul cargo, which flooded the Canadian market.[123] The glut forced salt prices to artificially low levels with which Ontario producers could seldom compete. In 1886 alone, six million barrels of English salt were imported into Canada. Ontario salt production that year was therefore limited to a mere three hundred thousand barrels, the bulk of which was sold as table and dairy salt in Ontario.

When the Salt Association failed in 1885, the domestic price of salt fell from 97 cents a barrel to its former price of approximately 54 cents.[124] As a result of this poor price and the market situation, only those salt operations with access to the purest brines and located close to cheap water transportation routes could afford to stay in business. And even then, most of those salt works operated only during the warm months, when the least amount of fuel was required for evaporation and the Great Lakes shipping season was open.[125] Thus, despite the numerous technological innovations developed locally to lower the costs of producing salt, production was confined to a handful of works located close to the lake and operating on a seasonal basis. The 1888 statistical returns for Ontario salt manufacturers indicate that the average salt block was capable of turning out 100 to 125 barrels of salt per day, yet the average annual production per block amounted to only one-tenth of that.[126]

In conclusion, neither the salt producers and processors nor the petroleum producers and refiners achieved the goals of large profits or dependable markets that should have been theirs. This was not for lack of technological borrowings or local inventions. The problem appears to have been the nature of the two resources, the physical setting in which they were found, and inability to compete economically with foreign producers, even for the domestic market.

7

TECHNOLOGICAL CHANGE IN A NEW AND DEVELOPING COUNTRY

The Ontario mining frontier was just one of hundreds of young regions affected by the demands of the industrial age for more and different minerals, and it looked to the new technological capacity of the time to make mining economically possible. This investigation demonstrates that mining techniques were often imported to Ontario, usually from Great Britain and the United States, but that they required considerable modification and improvement to be practical. And, in some important regards, new technology originated in Ontario, as seen in the case of the pioneer pertoleum industry of the western peninsula. Such technological adaptation and local invention was a key to the success of a frontier region in economic transition and, ultimately, to the perfection of existing techniques and the creation of new ones on a more international level.

In each mining district a few operations proved adept in trying out innovations, and these led the way in applying new technology adopted at their own and at other sites to attack the problems and costs associated with particular minerals and different settings. Not surprisingly, these leader operations tended to the largest, most heavily capitalized ones, carried out at the richest most accessible sites. They could afford to risk capital required to experiment with new technologies, for only they could benefit from the economies of scale, especially in the area of exploring for minerals. The exploring stage was especially risky, for finding an apparently workable deposit did not in itself guarantee a profitable operation. At the operating stage, the risks were different. As seen, mine operators were compelled to go deeper in the existing mine or well or work leaner or more complex materials. Either way, the costs would escalate. Even at the most promising operations innovations were introduced quite cautiously, usually incrementally, so as not to overtax capital resources or exceed the capacities of limited mineral reserves.

The case studies of the Bruce Mine, Silver Islet Mine, and the Dean and Williams Mine – all of which were active at different times and achieved different levels of economic success – indicate how widespread this pattern of incrementation was. Indeed, the introduction of sophisticated levels of technology to a mining operation prior to thorough testing of the deposit was an almost certain clue that the owners were not interested in mining ore but were trying to attract additional investment capital or dispose of their property at a profit. Such deceptions were very prevalent in the gold- and lead-mining operations of the southeastern district, as the story of the Gatling and Frontenac mines reveals.

Of course, some large areas of mining technology were never borrowed by mining operators because the techniques and machinery were overspecialized or inappropriate for the Ontario situation. Thus, the wide range of innovations inspired by coal-mining were not applied in Ontario because the industry was absent from the province. Another factor which blocked the adoption of new technologies at possibly suitable sites was the cost of replacing outmoded equipment, plants, and even the workforce itself. English mine owners, for example, tended to transfer to the province whole traditional systems of mining that depended on skilled workers as well as machinery and equipment. Hence, it was not always possible to introduce too much mechanical change without also changing the type of workers and contractual arrangements employed. And the owners were also committed to a previous capital investment in whatever mining system was in place. These factors help explain why the Bruce Mine, operating from the 1850s through the mid-1870s, appeared so little affected by the important technological developments that were occurring in the nearby Michigan copper-mining district.

Of greater importance, however, was the fact that on the frontier mining operations seldom developed beyond extracting high-grade ores, which in the initial phases were often more profitable to mine and market than lower-grade ores in more accessible areas. Because few mining operations extended to deep levels, compressed-air drills, hydraulic pumps, and sophisticated hoisting, hauling, and blasting systems were rarely introduced in this period. Moreover, many of the minerals being mined on the Ontario frontier – apatite, mica, and iron ore, for example – were being shipped out of the country in raw states. They were so low in value in relation to their bulk that the cost of transporting them was high. So they had to be mined as cheaply as possible. Only certain large-scale silver mines in the Upper Great Lakes district warranted adopting such equipment. Hand techniques, perhaps with some use of steam technology and simple concentration and by-product recovery techniques, were the aspects of mining that were the most commonly experimented with. And seemingly outmoded technology persisted in the face of new advances also in petroleum-drilling and pumping in Ontario's oilfields, which were shallow-lying and limited in extent. At the processing stage, ore roasting and smelting, petroleum refining, and salt evaporation all required considerable thermal energy. Since coal resources were lacking and local cordwood was becoming scarce, fuel conservation was an important and ever-present consideration in frontier mining that significantly influenced both the level and the type of processing technologies employed. Although an impressive amount of experimentation with various technologies had occurred in all of the districts, all in all, mining operators were limited in what they could borrow from the international pool of available technological innovation.

When operations in the province did try out foreign technology, they often became innovative, whether or not they actually adopted the technology in question. The borrowing of technology always entailed some degree of experimentation, adaptive change, and perhaps improvement of existing techniques. At the very least, the equipment would have had to be modified and its components arranged to suit the particular minerals being developed in each setting. Much of this activity was conducted on a modest scale, as in the case of experimentations with brine evaporation equipment and fuel-efficiency techniques in the western peninsula district and with gold-recovery processes in the southeastern district and the Lake of the Woods region. In certain cases, however, the improvements were substantial enough to constitute significant Canadian improvements to the regional, and even the international, pool of technology – the development of the "Canadian System" of petroleum drilling, introduced in the 1870s at Petrolia, the Frue Vanner at Silver Islet in the same decade, and the Frasch process of refining and deodorizing petroleum, for example. In all cases, local inventions for mining were directed to the exigencies of a newly developed area or toward exploiting the new field of petroleum.

In order to truly assess technological innovation on the frontier, it is necessary to identify the type of technological problems that occurred in marginal economies, the solutions that were sought for these problems, and, finally, those who were responsible for the innovations. Seldom did entrepreneurs or workers keep records of their technological activities – the diary of Thomas Nightengale and the letterbooks of William B. Frue and the Macdonnell and Foster patent agency are rare surviving examples. But one can study inventive activity by looking at patents. Patents in the nineteenth century were awarded for an art, machine, manufacture, composition of matter, or new and useful improvement thereof, for something original, technically feasible, and more than a minor adjustment. Patented inventions likely represent only a small fraction of the innovations actually developed for mining in the province and items that were patented were not necessarily used. Nevertheless, a systematic study of patents granted to Ontario residents for mining provides valuable insights into the process of local invention on the Ontario frontier.

TABLE 4
OCCUPATIONAL DISTRIBUTION OF ONTARIO PATENTEES FOR MINING INVENTIONS, 1861-1890

	I *Extracting & Treating Ore*	*IIa* *Petroleum* *(a) Refining etc.*	*IIb* *(b) Pumping Storing*	*III* *Salt*	*TOTALS*
1. Petroleum Refiner	0	13	1	0	14
2. Physician/M.D.	0	1	7	2	10
3. Gentleman	2	5	1	1	9
4. Machinist	4	2	3	0	9
5. Chemist	1	4	0	2	7
6. Merchant	1	1	1	2	5
7. Banker	0	0	0	5	5
8. Oil Producer	0	1	3	0	4
9. Oil Operator	0	1	1	1	3
10. Mine Manager	3	0	0	0	3
11. Miner	3	0	0	0	3
12. Salt Manufacturer	0	0	0	3	3
13. Farmer	0	0	0	3	3
14. Assayer	2	0	0	0	2
15. Manufacturer	0	1	0	1	2
16. Mechanic	0	0	2	0	2
17. Boilermaker	0	1	0	0	1
18. Millwright	1	0	0	0	1
19. Newspaper Publisher	0	1	0	0	1
20. Petroleum Merchant	0	1	0	0	1
21. Innkeeper	1	0	0	0	1
22. Carpenter	0	0	1	0	1
23. Tinsmith	0	0	1	0	1
24. Miller	0	0	0	1	1
25. Stationer	0	0	0	1	1
26. Magistrate	0	0	0	1	1
27. Yeoman	0	1	0	0	1
28. Engineer	1	0	0	0	1
Unspecified	5	12	7	4	28
TOTALS	24	45	28	27	124*

*Because in a few cases there are co-applicants for a single patent, the total number of patentees (124) slightly exceeds the total number of patents (117).

Technological experimentation and local invention were critical activities for members of the frontier community. Some patentees were directly involved in mining and primary processing, and others were called upon to solve specific technological problems. Some individuals simply had a general vested interest in the success of any local economic venture.

Before 1890, Ontario residents were granted well over a hundred patents expressly concerned with mining, mainly for primary processing and by-product recovery, and to overcome the particular problems of petroleum, such as safe storage. Approximately 59 per cent of these applied to the distilling, refining, storing, and deodorizing of petroleum, another 22 per cent applied to salt evaporation, while the remaining 19 per cent applied to a miscellany of improvements related to extracting, and especially to treating, the recalcitrant minerals of the Canadian Shield region. Roughly a hundred other patents were examined, but they are left out of consideration in this study because they were either too far removed from the stages of extraction and primary processing (for example, oil lamps and processes for manufacturing gas from oil) or not unequivocally related to mining (for example, rock drills and well-drilling apparatus). No evidence has surfaced to indicate that there were any major breakthroughs. Some patents entailed minor mechanical modifications of chemical or thermal processes pertinent to petroleum refining or fuel conservation in primary processing of ores. All the local inventors were working to perfect a method in a long line of development, rather than to create an entirely new technology. The patentees were not professional inventors. They were men of varying occupations, including miners, blacksmiths, mechanics, plumbers, machinists, chemists, druggists, doctors, manufacturers, and "gentlemen," who lived mainly in the mining districts and were, by and large, very familiar with the problems addressed in their patents.

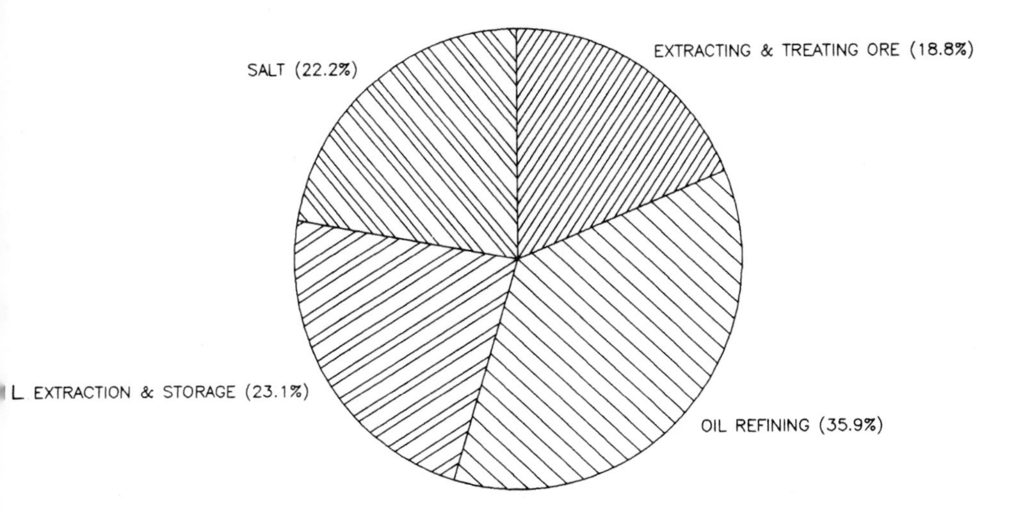

Figure 5. Ontario Patents by Mining Type, 1861-1890. *The borrowing of technology always entailed some degree of experimentation, even local invention. Before 1891, Ontario residents were granted well over a hundred patents expressly concerned with mining. These were mainly for primary processing and by-product recovery, and to overcome the particular problems of petroleum. In all cases, local inventions for mining on the Ontario frontier were directed to the exigencies of a newly developed region or toward exploiting the new field of petroleum.*

These early attempts to improve technology in Ontario provide an opportunity to speculate about the relationship between invention and transitional economic development in small, open societies. Clearly, several sectors of the community were involved in domestic invention, unlike the present pattern of corporate research and development. Ontario was caught up in the search for exploitable, exportable resources. Technological experimentation and local invention were, therefore, critical activities for members of the local community, for those who were directly involved in mining and primary processing, as well as for tradesmen and workers called on to solve specific technological problems. Others, such as physicians, druggists, and innkeepers, had a more general vested interest in the success of any domestic economic venture. They were in a position to keep abreast of local developments and to make their own contributions to technology.

Once innovations had been introduced to a district their diffusion on a regional basis was not always rapid or widespread, for lengthy trial periods were apt to be needed to develop and improve unique or complex technological innovations and to operate them, and special repair and maintenance facilities to keep them operating were required. The diamond-tipped exploration drill is a case in point. Simpler technology based on fairly traditional techniques, like the Frue Vanner, the jerker system for transmitting power to petroleum pumps, and steam-heated brine evaporators, were more quickly spread throughout the district. A special factor helping the rapid diffusion of petroleum and salt innovations was the regular or homogeneous nature of the two minerals. In either branch of the industry producers and manufacturers were handling essentially the same product and facing the same difficulties. Once a particular solution proved effective, it was likely to be applied throughout the industry. This was in sharp contrast with solid minerals as the quality, character, and mode of occurrence of the ores varied considerably from locale to locale, from mine to mine within a single locale, and even from level to level within a single opening. Consequently, the technological solutions were likely to be very specific and custom-designed to suit an individual mining operation.

The patent record and official mining reports reflect these conditions as regards the western peninsula and eastern regions. They demonstrate that the western peninsula was the most locally experimental, innovative mining region. The novelty as well as the unusual properties of petroleum (derived from hundreds of wells) called for a host of technological innovations to reduce the costs of drilling and pumping; overcome the problems of safe storage and transportation; of refining and by-product recovery; and developing new uses to enhance the commercial potential of the mineral fuel. Far behind was the southeastern district, where great numbers of generally small, individually unique, marginal operations were carried on over more longer periods of time than the size and character of the deposits warranted. The northern region relied almost exclusively on high technology imported from the older mining districts of Great Britain and the United States and engaged in very little inventive activity.

The government's longstanding relaxed attitude toward development of the province's mining operations evoked strong criticism at the time from mining men, investors, and politicians. These men were interested in having the provincial government assert control over systematic resource development in the province, and in starting to develop the resources of the vast northern territory that was opening up to development by construction of the trans-continental railway section across the territory in the 1880s. Despite the absence of a provincial bureau of mines and school of mines before 1890, however, the role of the dominion government in assisting and expanding provincial mining operations had been varied, often indirect, but nevertheless significant. The most outstanding contribution was that provided by the Geological Survey of Canada, which from the beginning assisted in mineral discovery and spread word of new technology. It accomplished this through reports of its members' field work, research and analyses, and a proliferation of descriptive and analytical articles by its officers. They were based on Canadian conditions and experience, and so were especially useful for Ontario mining operations. It offered a major incentive to full-scale mineral development – hence technological adoption and innovation – by subsidizing local railways through mineral lands, expecially in the southeastern iron belt. In general, the dominion government encouraged innovations in mining technology by means of the patent system, which offered inventors a short-term monopoly on their ideas. More specifically, the departments of inland revenue and marine and fisheries undertook scientific and technological research and product regulation in the area of petroleum.

To conclude, Canada today remains essentially a producer of staple exports with a large, primary manufacturing sector engaged in the rough finishing of resources for a narrow range of export markets. Dependence upon foreign markets and prices has greatly influenced the devlopmental patterns of industries such as mining and, in turn, the levels and types of technology employed. In the period under study, for example, the world price of mining outputs was low; therefore, Ontario production levels were low, and so were the levels of technology employed. Except for petroleum and salt, the minerals soon were shipped away from the country in semi-raw states to suit the needs and purposes of foreign markets. Petroleum and salt were different in two respects: They were restricted to the domestic market, and they required to be fully processed in Canada. Consequently, domestic technological innovation in these branches of the industry advanced into the refining and distilling stages, for example, to purifying illuminating oils from the crude petroleum, and finding uses for other by-product components, especially as cheap sources of fuel. And the country has continued to call forth innovations that may be needed for exploiting a new Canadian field of mining development, as in the case of nickel at the end of the nineteenth century and, more recently, uranium in the twentieth century. In both cases, the research took place outside Canada.

In terms of the general writing on the worldwide phenomena of technological change, the findings of this study seem to corroborate most of the work based on other cases, times, and settings. This is especially true for the importance of communications and scale factors, changing relative factor prices, the host of small improvements and inventions necessary to bring new techniques into practical use, fuel-economy and efficient power transmission aspects, and the role of learning-by-doing and leader operations in introducing and diffusing technological innovations within an industry and across a region. Most studies of technological change focus on single aspects, such as invention, diffusion, adaptation, or lag. They have the advantage of isolating a stage in the process of technological change for close examination, but they fall down by not viewing the complete process as a dynamic one.

In the present study I have chosen to examine technological change as a sequence of events and to apply this procedure to the broad, complex industry that was taking place in three highly varied social and physical environments and immensely different locales on the frontier. Doing this has made it possible to examine a wide variety of experimentations and borrowings ranging from the failure to discover a gold-recovery process in the Marmora-Madoc area, to the very modest distinctive contributions of salt manufacturers to the quite important technologically advanced operations taking place in the Thunder Bay silver district and in the Sudbury Basin. Such inquiry sheds new insights into the uneven quality of technological change in the context of a new and developing region, like the Ontario mining frontier. New technology was adopted whenever it was practical to do so. And technological innovations were developed locally where possible and necessary. In all cases, the most successful technological innovations – whether borrowed or developed within the province – were those that were best suited to the newness or frontier nature of mining operations in nineteenth-century Ontario. A well-suited technique was typically cast on a small scale, tailored to local sources of fuel, built from cheap materials available locally – notably wood – and easy to maintain. The broad geographical base adopted by this study of frontier mining also suggests the importance of examining the material environment – in the case of mining, the mode of occurrence and character, and the state of present knowledge of, various minerals being worked, and the organization of work – in order to truly understand the processes involved in technological change related to resource industries. All in all, this analysis of technological change on the Ontario mining frontier is relevant for studies of similar facets of other newly opening regions and countries, and to the better understanding of the actual process of innovation and diffusion that underlies specific technological changes.

APPENDIX

TABLE A1
TOTALS FOR NUMBERS OF MINERS AND RAW MINERAL PRODUCTS MINED IN ONTARIO, 1870/71-1890/91

	1870/71	*1880/81*	*1887*	*1890/91*
Total number of:				
Miners	267	493	–	1,034
Oil Refinery employees	–	379	–	–
Oil-well employees	–	–	–	155
Salt works employees	–	–	–	158
Total Raw Mineral Products:				
Silver	69,197 T	87,000 T	190,495 T	14,925 T
Gold	199 oz.	152 oz.	450 oz.	prospects
Copper Ore	1,934 T	170 T	567 T	prospects
Iron Ore	30,726 T	91,877 T	16,592 T	prospects
"Other" Ore	1,121 T	121 T	–	–
Crude Petroleum	12,969,435 gal.	15,490,662 gal.	26,737,655 gal.	894,647 bbl.
Salt	–	472,000 bbl.	429,807 bbl.	55,167 T

Sources: *Census of Canada, 1870-71,* vol. 2, table 13; vol. 3 table 27 (Ottawa: I.B. Taylor, 1875): *Census of Canada 1880-81,* vol. 2, table 14; vol. 3, tables 28 and 44 (Ottawa: MacLean, Roger, 1883): *Census of Canada, 1890-91,* vol. 2, table 12 (Ottawa: S.E. Dawson, 1893); Ontario Bureau of Mines, *Report, 1891* (Toronto: Warwick, 1892), p. 6; and *Ontario Commission on Mineral Resources* (Toronto: Warwick, 1890), p. 210.

TABLE A2
CANADIAN (ONTARIO) CRUDE PETROLEUM PRODUCTION AND EXPORTS, 1867-1891

	Canadian (Ontario) Production			*Canadian (Ontario) Exports*			
Year	*Quantity (000's bbls.)*	*Value (000's $)*	*Price per bbl. ($)*	*Quantity (000's bbl.s)*	*Value (000's $)*	*Price per bbl. ($)*	*Per cent of Total Production Exported*
1891	775	1,010	1.30	13	18	1.39	1.68
1890	795	903	1.14	12	18	1.50	1.51
1889	705	654	.93	7	11	1.58	.99
1888	695	714	1.03	6	75	12.50	.86
1887	714	557	.78	14	14	1.00	1.96
1886	584	526	.90	7	10	1.43	1.20
1885	588	–	–	10	11	1.10	1.70
1884	571	–	–	31	30	.97	5.43
1883	473	–	–	–	–	–	–
1882	390	–	–	–	–	–	–
1881	369	–	–	–	–	–	–
1880	350	–	–	–	–	–	–
1879	575	–	–	–	–	–	–
1878	312	–	–	–	–	–	–
1877	312	–	–	–	–	–	–
1876	312	–	–	–	–	–	–
1875	220	–	–	–	–	–	–
1874	169	–	–	–	–	–	–
1873	365	–	–	–	–	–	–
1872	308	–	–	–	–	–	–
1871	270	–	–	–	–	–	–
1870	250	–	–	–	–	–	–
1869	220	–	–	–	–	–	–
1868	200	–	–	–	–	–	–
1867	190	–	–	–	–	–	–

Source: M. C. Urquhart, ed., *Historical Statistics of Canada* (Toronto: Macmillan, 1965), Series N170-175, pp. 437-38; N182-187, pp. 440-441. Exports are clearly nominal. (We may assume that the western peninsula district of Ontario is responsible for all crude petroleum production in the dominion.)

TABLE A3
CANADIAN PETROLEUM MANUFACTURED AND EXPORTED, 1868-1877*

Year	*Dominion Total Manufactured and Dutiable (gallons)*	*Total Outside Ontario Manufactured and Dutiable (gallons)*	*Per cent of Dominion Total Manufactured and Dutiable Outside Ontario*	*Manufactured and Exported (gallons)*	*Manufactured for Use in Dominion Lighthouses (Tax-free) (gallons)*
1877	7,913,754	8,274	.10	–	–
1876	4,838,215	1,823	.04	–	–
1875	4,811,596	33,128	.69	1,140	21,687
1874	6,752,282	10,362	.15	1,065,787	58,147
1873	14,602,087	81,891	.56	9,597,525	73,786
1872	12,323,991	114,808	.93	8,076,885	64,364
1871	11,689,761	61,332	.52	5,531,780	77,238
1870	10,736,636	1,516	.01	–	20,014
1869	2,772,224	3,088	.11	–	N/A
1868	237,765	0	0	–	N/A

Source: Canada, Department of Inland Revenue, *Annual Reports, Returns and Statistics of the Inland Revenues of the Dominion of Canada,* for 1870-1891.

All the Canadian refineries were located in Ontario and Quebec (Montreal only), but the above indicates how insignificant were the Quebec operations in terms of total production. The figures for the first years, 1868-69, are low because they do not take into account stocks produced before the excise duty came into effect in April 1868. 1873 marks the year in which Standard Oil Trust was formed in the United States. Note how domestic manufacturing of petroleum and its export drops off sharply for the next few years. *In 1877 the Excise duty on petroleum manufacture was repealed (Canada, *Statutes,* 40 Vict. chap. 14, 20 February 1877). Thereafter the Department of Inland Revenue continued to inspect refineries but no longer reported the quantities produced in Canada or exported.

TABLE A4
CANADIAN (ONTARIO) SALT PRODUCTION AND EXPORTS, 1875-1891

	Canadian (Ontario) Production			*Canadian (Ontario) Exports*			
Year	*Quantity (000's CWT)*	*Value (000's $)*	*Price per CWT ($)*	*Quantity (000's CWT)*	*Value (000's $)*	*Price per CWT ($)*	*Per cent of Total Production Exported*
1891	900	161	.18	4	1	.25	.44
1890	880	199	.23	5	1	.20	.57
1889	660	130	.70	6	2	.33	.90
1888	1,180	185	.16	11	4	.36	.93
1887	1,200	166	.14	108	12	.11	9.00
1886	1,240	227	.18	157	17	.10	12.66
1885	–	–	–	173	19	.11	–
1884	–	–	–	117	15	.13	–
1883	–	–	–	140	19	.14	–
1882	–	–	–	127	18	.14	–
1881	–	–	–	240	45	.19	–
1880	–	–	–	327	46	.14	–
1879	–	–	–	414	49	.12	–
1878	–	–	–	285	37	.13	–
1877	–	–	–	493	61	.12	–
1876	–	–	–	637	84	.13	–
1875	–	–	–	380	67	.18	–

Source: M.C. Urquhart, ed., *Historical Statistics of Canada* (Toronto: Macmillan, 1965), Series N89-119, p. 427; N143-162, p. 435. (We can assume that all commercial salt production in the dominion takes place in the western peninsula district of Ontario. One barrel of salt weighed 280 lbs.).

NOTES

NOTES TO CHAPTER ONE

1. William M. Marr and Donald G. Paterson, *Canada: An Economic History* (Toronto: Macmillan, 1980), p. 31.
2. For copper supplies and prices, see W.B. Gates, Jr., *Michigan Copper and Boston Dollars: An Economic History of the Michigan Copper Industry* (Cambridge, Mass.: Harvard University Press, 1951), pp. 6-7, 44-45; and O.C. Herfindahl, *Copper Costs and Prices, 1870-1957* (Baltimore: Johns Hopkins University Press, 1959), pp. 70-71. See also K.W. Taylor and H. Mitchell, *Statistical Contributions to Canadian Economic History*, vol. 2 (Toronto: Macmillan, 1931).
3. Marr and Paterson, *Canada: An Economic History*, pp. 303-4, 358-60; U.S., House of Representatives, Subcommittee on Natural Resources, *Technological Trends and National Policy, Including the Social Implications of New Inventions* (Washington: USGPO, 1937), pp. 145-76; Homer Aschman, "The Natural History of a Mine," *Economic Geography* 46 (1970):172-89; and Geoffrey Blainey, "A Theory of Mineral Discovery: Australia in the Nineteenth Century," *Economic History Review* 2d ser., 23, 2 (1970):298-313.
4. See Joseph Schumpeter's classic works, *The Theory of Economic Development* (Cambridge, Mass.: Harvard University Press, 1934); and *Business Cycles: A Theoretical, Historical, and Statistical Analysis of the Capitalist Process* (New York: McGraw-Hill, 1934).
5. See Alfred D. Chandler, Jr., "Anthracite Coal and the Beginnings of the Industrial Revolution in the United States," *Business History Review* 46 (Summer 1972):143-81.
6. See Jacob Schmookler, "The Interpretation of Patent Statistics," *Journal of the Patent Office Society* [*JPOS*] 32, 2 (1950):123-46; "The Utility of Patent Statistics," *JPOS* 35, 6 (1953): 407-12; and "Patent Office Statistics as an Index of Inventive Activity," *JPOS* 35, 8 (1953):539-50.
7. Richard R. Nelson, "Introduction," in *The Rate and Direction of Inventive Activity*, Report of the National Bureau of Economic Research, No. 13 (Princeton: Princeton University Press, 1962), p. 5. See also Edwin T. Layton, Jr., "Mirror Image Twins: The Communities of Science and Technology in 19th Century America," *Technology and Culture* 12, 4 (1971):562-80.
8. Richard H. Shallenberg, "Evolution, Adaptation and Survival: The Very Slow Death of the American Charcoal Iron Industry," *Annals of Science* 32, 4 (July 1975):341-58.
9. Nathan Rosenberg, "Technological Change in the Machine Tool Industry, 1840-1910," *Journal of Economic History* 23 (1963):414-43; Carroll W. Pursell, Jr., *Early Stationary Steam Engines in America: A Study in the Migration of a Technology* (Washington: Smithsonian Institution Press, 1969); Peter Temin, "Steam and Waterpower in the Early Nineteenth Century," *Journal of Economic History* 26 (1966):187-205; Larry D. Lankton, "The Machine *Under* the Garden: Rock Drills Arrive at the Lake Superior Copper Mines, 1868-1883," *Technology and Culture* 24, 1 (1983):1-37; and Jennifer Tann, "Fuel-Saving in the Process Industries during the Industrial Revolution: A Study in Technological Diffusion," *Business History* 15 (July 1975):149-59.
10. Louis C. Hunter, *A History of Industrial Power in the United States, 1780-1930*, vol. 1, *Water Power in the Century of the Steam Engine* (Charlottesville: University Press of Virginia, 1979).
11. Shallenberg, "Evolution, Adaptation and Survival," pp. 341-58; and Bruce E. Seely, "Blast Furnace Technology in the Mid-19th Cen-

tury: A Case Study of the Adirondack Iron and Steel Company," *IA* 7, 1 (1981):27-54.

12. Douglass C. North and Robert Paul Thomas, *The Rise of the Western World: A New Economic History* (Cambridge: The University Press, 1973), pp. 5ff. See also Edward Mansfield, "Technical Change and the Rate of Imitation," *Econometrica* (1961):741-66, which is a standard work on the subject.
13. Paul A. David, *Technical Choice, Innovation and Economic Growth: Essays on American and British Experience in the Nineteenth Century* (Cambridge: Cambridge University Press, 1975); and Robert C. Allen, "Collective Invention," *Journal of Economic Behaviour and Organization* 4 (1983):1-23.
14. For a discussion of the shift in source of inventive effort, see Carolyn Shaw Solo, "Innovation in the Capitalist Process: A Critique of the Schumpeterian Theory," *Quarterly Journal of Economics* 65, 3 (1951):417-28; and, more recently, Allan R. Pred, *Spatial Dynamics of U.S. Urban-Industrial Growth, 1800-1914: Interpretive and Theoretical Essays* (Cambridge, Mass,: M.I.T. Press, 1966), p. 87 n 3; and Wilbur R. Thompson, "Locational Differences in Inventive Effort and Their Determinants," in *The Rate and Direction of Inventive Activity,* Report of the National Bureau of Economic Research, No. 13 (Princeton: Princeton University Press, 1962), pp. 257ff.
15. Torsten Hägerstrand, *Innovation Diffusion as a Spatial Process,* Postscript and Translation by Allan Pred (Chicago: University of Chicago Press, 1967); and see Allan R. Pred, "Industrialism, Initial Advantage, and American Metropolitan Growth," in *Geographic Perspectives on America's Past: Readings on the Historical Geography of the United States,* ed. David Ward (New York: Oxford University Press, 1979), pp. 279-82.
16. H.J. Habakkuk, *American and British Technology in the Nineteenth Century: The Search for Labour-Saving Inventions.* (Cambridge: The University Press, 1962). Habakkuk's influence on later scholars is evident in S.B. Saul, ed., *Technological Change: The United States and Britain in the Nineteenth Century,* Debates in Economic History, Peter Mathias, ed. (London: Methuen, 1970); and Paul Uselding and Bruce Juba, "Biased Technical Progress in American Manufacturing," *Explorations in Economic History* 11 (1973):55-72.
17. David, *Technical Choice.*
18. See, for example, Robert H. Wiebe, *The Search for Order, 1877-1920* (New York, 1967); and David F. Noble, *America by Design: Science, Technology, and the Rise of Corporate Capitalism* (New York: Knopf, 1977).
19. Gates, *Michigan Copper and Boston Dollars,* p. 58; and Lankton, "The Machine *Under* the Garden," p. 4.
20. Lawrence A. Brown, *Innovation Diffusion: A New Perspective* (New York: Methuen, 1981), pp. 183-84.
21. See Douglass C. North, "Location Theory and Regional Economic Growth," *Journal of Political Economy* 63 (June 1955):248; and also, Melville Watkins, "Technology in Our Past and Present," in *Canada: A Guide to the Peaceable Kingdom,* ed. by William M. Kilbourne (Toronto: Macmillan, 1970), pp. 285-92. North drew heavily upon the Canadian work of Harold Innis.
22. Shallenberg, "Evolution, Adaptation and Survival;" Gates, *Michigan Copper and Boston Dollars;* and Harold A. Innis, *Settlement and the Mining Frontier,* Canadian Frontiers of Settlement, eds. W.A. MacKintosh and W.L.G. Joerg, vol. 9. 1936 (Millwood, N.Y.: Kraus Reprint, 1974).
23. For an historiographical overview, see Carl Berger, *The Writing of Canadian History. Aspects of English Canadian Historial Writing: 1900 to 1970* (Toronto: University of Toronto Press, 1976), chapter 4. The usefulness of the staples thesis to economic history is discussed in Marr and Paterson, *Canada: An Economic History,* pp. 10-18.
24. O.J. Firestone, *Canada's Economic Development 1867-1953 with Special References to Changes in the Country's National Product and National Wealth,* International Association for Research in Income and Wealth, ser. 7 (London: Bowes and Bowes, 1958), p. 5; and Taylor and Mitchell, *Statistical Contributions to Canadian Economic History,* vol. 2.
25. Sally F. Zerker, "Centralization in the International Oil Industry" (paper read before the Annual Meeting of the Canadian Historical Association, Vancouver 6-8 June 1983), p. 19.
26. Summarized in ibid, pp. 20-24. See also Centennial Seminar on the History of the Petroleum Industry, *Oil's First Century* (Cambridge, Mass.: Harvard University Press, 1959); Ralph W. Hindy and Muriel E. Hindy, *Pioneering in Big Business 1882-1911* (New York: Harper, 1955); Allan Nevins, *John D. Rockefeller: The Heroic Age of American Enterprise,* vol. 1 (New York: Scribner's, 1940); Ida

M. Tarbell, *The History of Standard Oil Company* (New York: Norton, 1969): and Harold F. Williamson and Arnold R. Daum, *The American Petroleum Industry*, vol. 1, *The Age of Illumination, 1859-1899* (Evanston: Northwestern University Press, 1959).

27. See Taylor and Mitchell, *Statistical Contributions to Canadian Economic History*, vol. 2, pp. 34-39; Gates, *Michigan Copper and Boston Dollars*, pp. 44-45; and Herfindahl, *Copper Costs and Prices, 1870-1957*, pp. 70-71. In the years between 1850 and 1880, the Michigan mines accounted for between 50 to 85 per cent of American domestic outputs of copper.
28. See Berger, *The Writing of Canadian History*, chapters 4, 5, and 6 on Innis and his followers.
29. Albert Faucher, *Québec en Amérique au XIXe siècle: essai sur les charactèrs économiques à la Laurentie* (Montreal: Fides, 1973); and Albert Faucher, "The Decline of Shipbuilding at Quebec in the Nineteenth Century," *Canadian Journal of Economics and Political Science* 23, 2 (1957):195-215. Morris Zaslow, "The Frontier Hypothesis in Recent Historiography," *Canadian Historical Review* 29 (1948): 153-66. Zaslow applies the notion of cultural transfer to frontier areas in his article, "The Yukon: Northern Development in a Canadian-American Context," in *Regionalism in the Canadian Community, 1867-1967*, ed. Mason Wade (Toronto: University of Toronto Press, 1969), pp. 180-97. Kenneth Kelly combines those notions of cultural transfer and environmental factors to suggest that a "duality of development" operated in the case of farming methods in pioneer Ontario. His research further reveals that local people were more important that foreign publications in determining what farming methods were most suitable. Kenneth Kelly, "The Transfer of British Ideas on Improved Farming in Ontario during the First Half of the Nineteenth Century," *Ontario History* 63 (1971): 103-11. Watkins, "Technology in our Past and Present;" J.M.S. Careless, "Frontierism, Metropolitanism, and Canadian History," *Canadian Historical Review* 35 (1954):1-21; and A.R.M. Lower, *Great Britain's Woodyard: British America and the Timber Trade, 1763-1867* (Montreal: McGill-Queen's University Press, 1973).
30. See, for example, J.J. Brown, *Ideas in Exile: A History of Canadian Invention* (Toronto: McClelland and Stewart, 1967); and Norman R. Ball's "Canadian Petroleum Technology during the 1860's" (Master's thesis, University of Toronto, 1972). But O.J. Firestone is one of the few scholars who has actually analysed the Canadian patent record. O.J. Firestone, *Canada's Economic Development 1867-1953;* and "Innovations and Economic Development — The Canadian Case," *Review of Income and Wealth* ser. 18, 8 (1972):399-419.
31. Elwood S. Moore, *American Influence in Canadian Mining*, Political Economy Series No. 9 (Toronto: University of Toronto Press, 1941).
32. Bruce Sinclair, N.R. Ball, and O.J. Petersen, eds., *Let Us Be Honest and Modest: Technology and Society in Canadian History* (Toronto: Oxford University Press, 1974); James Otto Petersen, "The Origins of Canadian Gold Mining: The Part Played by Labour in the Transition from Tool Production to Machine Production" (Ph.D. dissertation, University of Toronto, 1977); Duncan A. Stacey, "Technological Change in the Fraser River Salmon Canning Industry, 1871-1912" (Master's thesis, University of British Columbia, 1978); Allan Skeoch, "Technology and Change in Nineteenth-Century Ontario Agriculture" (Master's thesis, University of Toronto, 1976); Donald MacLeod, "Mines, Mining Men, and Mining Reformers: Changing the Technology of Nova Scotian Gold Mines and Colleries, 1858 to 1910" (Ph.D. dissertation, University of Toronto, 1981); Robert E. Ankli, H. Dan Helsberg, and John Herd Thompson, "The Adoption of the Gasoline Tractor in Western Canada," in *Canadian Papers in Rural History*, ed. Donald H. Akenson, vol. 2 (Gananoque: Langdale Press, 1980), pp. 7-39; and Richard Pomfret, "Mechanization of Reaping in Nineteenth Century Ontario: A Case Study of the Pace and Causes of the Diffusion of Embodied Technical Change," *Journal of Economic History* 36 (June 1976):399-415.
33. Ibid.; Faucher, "The Decline of Shipbuilding at Quebec in the Nineteenth Century"; and C.K. Harley, "On the Persistence of Old Techniques: The Case of North American Wooden Shipbuilding," *Journal of Economic History* 33, 2 (1973):372-98.
34. This procedure is based on the material mentioned in this chapter and is an expanded version of Dario Menanteau-Horta's model for interpreting the diffusion and adoption of new agricultural practices among Chilean farmers. With the Menanteau-Horta model, no local innovation is taken into

account. Dario Menanteau-Horta, "Diffusion and Adoption of Agricultural Practices among Chilean Farmers: A Sociological Study of the Processes of Communication and Acceptance of Innovations as Factors Related to Social Change and Agricultural Development in Chile" (Ph.D. dissertation, University of Minnesota, 1967), pp. 38-44.

35. This category of mineral operation is thoroughly documented by a geographer, Alexander Marshall Blair, in "Surface Extraction of Non-Metallic Minerals in Ontario Southwest of the Frontenac Axis" (Ph.D. dissertation, University of Illinois, 1965). His work focuses on the twentieth century. It examines the factors behind the changing location of extraction site and addresses the future problems for resource managers in Ontario.

NOTES TO CHAPTER TWO

1. Raphael Samuel, "The Workshop of the World: Steam Power and Hand Technology in Mid-Victorian Britain," *History Workshop Journal* 3 (1977):6-72, provides an interesting demonstration of the "spotty" nature of mechanization during the nineteenth century. For a discussion of the shift in the source of inventive effort, see Carolyn Shaw Solo, "Innovation in the Capitalist Process: A Critique of the Schumpeterian Theory," *Quarterly Journal of Economics* 65, 3 (1951): 417-28. The seminal work on the interindustry transferability of technological innovations in problem solutions is Nathan Rosenberg, "Technological Change in the Machine Tool Industry, 1840-1910," *Journal of Economic History* 23 (1963):414-43.
2. The standard work on this is Georgius Agricola, *De re Metallica* (New York, 1950). See also Cyril Stanley Smith, "Metallurgy in the Seventeenth and Eighteenth Centuries," in *Technology in Western Civilization,* vol. 1, eds. Melvin Kransberg and Carroll W. Pursell, Jr. (New York: Oxford University Press, 1967), pp. 143-46.
3. For an excellent overview, see A.B. Parsons, ed., *Seventy-Five Years of Progress in the Mineral Industry, 1871-1946* (New York: AIME, 1947). Alfred D. Chandler, Jr., provides a thoughtful discussion of the impact of mechanization on industry. See *The Visible Hand: The Managerial Revolution in American Business* (Cambridge, Mass.: Belknap Press, 1977), especially part 1.
4. Morris Zaslow, *Reading the Rocks: The Story of the Geological Survey of Canada, 1842-1972* (Toronto: Macmillan, 1975), p. 36.
5. The best examples of early experiments with the principles employed in the diamond drill are Beart's Improved Boring Apparatus, developed in Manchester, England, and patented in 1844; and the Fauvelle System of Boring, developed in France in 1846. See *Mechanic's Magazine* 42, 1123 (1845):111; and *Transactions of the North-of-England Institute of Mining Engineers* [NEIME *Trans*] 2 (1853-54):60. See also the history of fluid circulating drills in J.E. Brantly, *History of Oil Well Drilling* (Houston: Gulf Publishing, 1971), p. 259ff.
6. Brooman's Drilling Apparatus, patented in the United Kingdom in 1862.
7. "Leschot's Diamond-Pointed Steel Drill," *Scientific American,* 22, 18 (1870):282-83.
8. Ibid. In their improved version, Holt and Severance set the diamonds more securely and fastened them in a convex, rather than a concave, plane to allow greater clearance for the free descent of the drill and for the movement of the core up the tubular stem of the ring-cutter. They also raised the rotating speed to four times that of the original, constructed the feed to adjust automatically to the type of strata encountered, and added a swivel head for greater flexibility, all of which increased the ability of the drill to advance more quickly under difficult circumstances. They also replaced the fixed cylinder engine of the original arrangement by an oscillating cylinder engine.
9. De Volson Wood, "Drilling Machines at the Hoosac Tunnel," *Journal of the Franklin Institute* 3d. ser. 44 (August 1867):83-86; "Leschot's Diamond-Pointed Steel Drill"; A. Bassett, "The Diamond Drill," NEIME *Trans* 23 (1873-74):179-96; and Major Beaumont, "On Rock Boring by the Diamond Drill and Recent Application of the Process," *Proceedings of the Institution of Mechanical Engineers* [IME *Proc*] (April 1875):82-125.
10. Charles F. Jackson, "Metal Mining Practice over Sixty Years," *Canadian Mining Journal* 60, 11 (1939):674. See also L.A. Riley, "Cost and Results of Geological Explorations with the Diamond Drill in the Anthracite Region of Pennsylvania," *Transactions of the American Institute of Mining Engineers* [AIME *Trans*] 5 (1876-77):303; and Dan Fivehouse, *The Diamond Drill Industry* (Saanichton, B.C.: Hancock House Publishers, 1976).

11. "Report of Operations in Boring for Coal with the Diamond-Pointed Steam Drill . . . by R.W. Ells," in GSC, *Report of Progress for 1872-73* (1873), pp. 230-37 [hereafter, "Ells's Report, 1872-73"]; "Second Report on the Boring Operations with the Diamond Drill . . . by R.W. Ells," in GSC, *Report of Progress, 1874-75* (1876), pp. 90-93 [hereafter, "Ells's Second Report, 1874-75"]; and "Summary Report of the Director A.R.C. Selwyn. Appendix II, Report of Operations in Manitoba with the Diamond Pointed Steam Drill, 1873, by W.B. Waud," in GSC, *Report of Progress, 1873-74* (1874).
12. "Ells's Report, 1872-73"; and "Ells's Second Report, 1874-75."
13. The "Glossary of Terms," presented in Ontario, Royal Commission on Mineral Resources, *Report* (Toronto: Warwick and Sons, 1890) [hereafter, *Ontario Commission on Mineral Resources*], pp. 528-45, is fairly complete and has the advantage of explaining contemporary usages. For a detailed discussion of nineteenth-century underground operations (metallic mining) terminology, see Walker de Marquis Wyman, Jr., "The Underground Miner, 1860-1910: Labor and Industrial Change in the Northern Rockies" (Ph.D. dissertation, University of Washington, 1971), introduction.
14. John Richardson, "On the Application of Portable Engines for Mining Purposes," IME *Proc* (July 1873):167-201.
15. Smith, "Metallurgy in the Seventeenth and Eighteenth Centuries," pp. 143-46.
16. Robert M. Vogel, "Building in the Age of Steam," in *Building Early America*, ed. Charles E. Petersen (Radnor, Penn.: Chilton Book Co., 1976), pp. 133-34.
17. M.F.L. Cornet, "On the Application of Machines Worked by Compressed-Air in the Collieries . . . in Belgium [and England]," NEIME *Trans* 21 (1871-72): 199-220; and William Daniel, "On Compressed-Air Machinery for Underground Haulage," IME *Proc* (August 1874):204-33.
18. The first practical use of electricity in North American mining (other than for signalling and blasting) was at the Lyken Collieries, about 1887. There a colliery electrician named Schlesinger developed an electric locomotive and system of electric power transmission. See announcement in *Colliery Engineer* 8, 2 (1887):43-44; H.C. Spaulding, "Electric Power-Transmission in Mining Operations," AIME *Trans* 19 (1890-91): 258-88; Frank Brain, "Electrical Pumping in Collieries," *Colliery Engineer* 8, 6 (1888):126-27 (reprinted from the South Wales Society of Engineers, *Transactions*, n.d.); and S. Luxmore, "Early Applications of Electricity to Coal Mining," *Proceedings of the Institute of Electrical Engineers* 126 (September 1979): 869-74.

    Chenot's Improved Pneumatic Rock Drill, which was developed in France in 1882, was a machine for forming horizontal grooves in mines or quarries to permit rock to be taken out in blocks. It was electrically powered by a Gramme dynamo (*Scientific American* 48, 16 [1883]:23-24 (reprinted from *Publication Industrielle* ([1882]). See *Colliery Engineer* 8, 2 (1887):43-44.
19. See Robert M. Vogel, "Tunnel Engineering," Paper 41, USM Bulletin 240, *Contributions from the Museum of History and Technology* (1964):201-40.
20. Ibid., p. 209-10.
21. See Ibid., p. 210; Jackson, "Metal Mining Practice," p. 674; and E. Gybbon Spillsbury, "Rock-Drilling Machinery," AIME *Trans* 3 (1874-75):144-50. A steamdriven drill was patented (though not necessarily employed) in the United States in 1848; its owners claimed it possessed most of the desired virtues. See description of Foster and Bailey's Improved Machine for Drilling Rocks in *Scientific American* 3, 20 (1848):153. The Fowle's drill rod was attached directly to, and reciprocated by, a double-acting steam piston, and its operation was truly independent of gravity, permitting drilling in any direction.
22. Vogel, "Tunnel Engineering," pp. 210-12; Daniel, "On Compressed-Air Machinery for Underground Haulage," pp. 204-33; and Jackson, "Metal Mining Practice," pp. 673-74.
23. Wood, "Drilling Machines at the Hoosac Tunnel"; and Vogel, "Tunnel Engineering," p. 210.
24. Note 23 and *Scientific American* 22, 7 (1870): 106.
25. See details on the Beaumont Percussion Rock Drill (1864) in *Engineering* (6 November 1865):442-43; and NEIME *Trans* 18 (1868-69): Appendix 1. And see also Low's Rock Boring Machine (1865), in George Low, "Description of a Rock Boring machine," IME *Proc* (April 1865):179-200. The English had in fact experimented with a crude form of compressed-air drill, termed a "wind hammer," for coal-mining as early

as 1844 (C. Bruton, "Design of a Wind Hammer for Boring Rocks," *Transactions of the Royal Cornwall Polytechnic Institute* [1844]).

26. See, for example, the Ingersoll air drill, patented 1871, described in Spillsbury, "Rock Drilling Machinery;" and in *Canadian Mining Review* (September 1888): advertisement. The Rand "Little Giant," and other Rand drills are detailed in "American Industries No. 63. The Manufacture of Power Drills for Mining, Excavating, Ec," *Scientific American* 43, 21 (1880):399, 402; and George Koether, *The Building of Men, Machines and a Company* (Ingersoll-Rand Co., 1971). The German drill, "Sach" is described in *Engineering* 13 (22 March 1872):200, and 31 (12 November 1880):429, 434; and Thomas Jordon, "On Rock-Drilling Machinery," IME *Proc* (April 1874):77-100. Improvements to Sach's drill included connecting it to the Doring Frame. See also a description of a French drill, the Osterkamp Compressed Air Rock Drill, developed in 1873, in *Engineering* 15 (16 January 1873):395. Improvements to this drill involved the reciprocating movement of the piston. The drill was exhibited at the Vienna Exposition, 1873. Other improvements to stands and mounts for drills included three developed in 1874: the Jordon Tubular Stand, developed in England (Jordon, "On Rock-Drilling Machinery"); the Rand Rock Drill on Patent Tunnel or Drifting Column, patented in the U.S. (A.C. Rand, "A New Rock Drill without Air Cushion," AIME *Trans* 13 [*1884-85*]:249-53; and *Scientific American* 37, 21 [1877]:319-20); and the Waring Air Drill, also developed in the United States (Spilsbury, "Rock-Drilling Machinery," pp. 147-49). In the Waring, the working parts were covered and the drill was mounted on a tripod, attached by a single bolt. The legs were constructed so that the drill could be placed in any position.
27. Larry D. Lankton, "The Machine *Under* the Garden: Rock Drills Arrive at the Lake Superior Copper Mines, 1868-1883," *Technology and Culture* 24, 1 (1983): 1-37. Calumet and Hecla Mines employed sixty-five Rand drills by 1882.
28. Jordon, "On Rock-Drilling Machinery," pp. 77-100.
29. Spillsbury, "Rock-Drilling Machinery," pp. 147-9.
30. "Leschot's Diamond-Pointed Rock Drill," p. 283.
31. The Gardner Steam Rock Drill, developed in New York in 1853 was a small portable arrangement worked by hand or by steam. It was exhibited at the New York Crystal Palace, 1853 (*Scientific American* 8, 42 [1853]: 340). Bank's Automatic Rock Drill was developed in Connecticut and patented in 1869. All operating machinery was sustained upon an upright frame. The main shaft carried at one end a crank or pulley. At the other end, several hammers struck the head of the drill. The machine was operated by hand or power. *Scientific American* 21, 5 (1869):65.
32. Examples include Barlow's Handpowered Rock-Drilling Machine, developed in Manchester, England, in 1869 for use when neither steam nor compressed air was available. The tool was struck rapidly by a steel-faced hammer worked by a crank through the medium of a spring. In the United States, in 1888, the new Ingersoll Duplex Hand-Powered Rock Drill was advertised as being so simple to construct that any labourer could operate it. It automatically fed as it cut (*Mining World and Engineering Record* [3 March 1888]:292, 302).
33. An excellent discussion of these developments is contained in Otis E. Young, Jr., *Western Mining* (Norman: University of Oklahoma Press, 1970), pp. 187-88.
34. Discussed in a general way in Arthur M. Johnson, "Expansion of the Petroleum and Chemical Industries, 1880-1900," in *Technology in Western Civilization,* vol. 1, eds. Melvin Kransberg and Carroll W. Pursell, Jr. (New York: Oxford University Press, 1967), pp. 675-77. See also an invaluable source, U.S., 41st Congress, 2nd. Session, 1870, *Report of Rossiter W. Raymond on the Statistics of Mines and Mining in the States and Territories West of the Rocky Mountains,* "The Mechanical Appliances of Mining," (hereafter "Report of Rossiter Raymond on the Mechanical Appliances of Mining"), pp. 535-50.
35. Ibid., p. 489; and Vogel, "Tunnel Engineering," pp. 213-14.
36. Ibid., p. 213; Henry Sturgess Drinker, *Tunneling, Explosive Compounds and Rock Drills* (New York: Wiley, 1878): and W.H. Ellis, "Nitro-Glycerine: Its History, Manufacture, and Industrial Application," *Canadian Journal* n.s. 14, 4 (1875):356-66.
37. Vogel, "Tunnel Engineering," pp. 213-15. Under ideal conditions forty per cent more tunnel length was advanced per cycle of operations. "Mowbray's Frictional Powder

Keg Battery and Mica Blasting Powder" was patented in the U.S. in a series of fourteen separate patents issued between 1868 and 1875. See Wood, "Drilling Machines at the Hoosac Tunnel," pp. 9-10; and *Scientific American* 22, 7 (1870):106.

38. Most of these received a trial in the well-publicized project to clear the navigation channel at Hell's Gate, alongside Manhattan Island, New York (*Scientific American* 22, 7 [1870]:106). Also see "High Explosives: Dynamite or Giant Powder," *Scientific Canadian Mechanics' Magazine and Patent Record* 9 (August 1881):246-47.
39. For a detailed discussion, see Johnson, "Expansion of the Petroleum and Chemical Industries," pp. 676-77.
40. Nobel's Explosive Gelatine was patented in Canada 1876 (No. 6,869). See *Canadian Mechanics' Magazine* (6 September 1878):286; *Canadian Mining Journal* 2, 10 (1884):7; and *Colliery Engineer* (December 1887):110-11.
41. Allan G. Talbot, "Ontario Origin of the Canadian Explosives Industry," *Ontario History* 56 (1964):37-44. An excellent overview of the commercial application of nitroglycerine is contained in "High Explosives: Dynamite or Giant Powder."
42. Patented 8 April 1873, Canada Patent No. 2,242, and re-issued 6 April 1878, No. 8,624.
43. Agricola, *De re metallica.*
44. For a description of Cornish stamps, see A.K. Rouse, "Borlase's Cornwall," *History Today* 23, 12 (1973):879-83; and R.J. Frenchville, "The Results Obtained by the Cornish System of Dressing Tin Ores," *Transactions of the Mining Association and Institute of Cornwall* [MAIC *Trans*] 1, 2 (1886):93-104. The California version is explained in George W. Small, "Notes on the Stamp Mill and Chlorination Works of the Plymouth Consolidated Gold Mining Company, Amadore County, California," AIME *Trans* 15 (1886-87):305-8. For a complete overview of the nineteenth-century improvements in stamps, see Arthur F. Taggart, "Seventy-Five Years of Progress in Ore-Dressing," in *Seventy-Five Years of Progress in the Mineral Industry,* ed. A.B. Parsons, pp. 82-125, especially table 1, "Developments in Comminution."
45. See Gates, *Michigan Copper and Boston Dollars,* pp. 26, 68, n. 92; and E.J. Bennett, "Modern Stamping and Concentration Appliances," MAIC *Trans* 2, 2 (1889):56-66.
46. A detailed contemporary account is to be found in John F. Blandy, "Stamp Mills of Lake Superior," AIME *Trans* 2 (1873-74): 208-15; Frederick G. Cogging, "Notes on the Steam Stamp," *Engineering* 40 (January 1886):119, (February 1886):130-31, 201-3; and Edgar P. Rathbone, "On Copper Mining in the Lake Superior District," IME *Proc* (February 1887):86-123.
47. See *Canadian Mining Review* 9, 7 (July 1890):102; and *Canadian Patent Office Record* 14 (1888).
48. See notes on individual washers and amalgamators in various issues of the *Scientific American* 8 (13 November 1852):65 (Berdan's stamps); 1 (29 October 1859):296 (Murry's Stamping Machine); and 15 (14 July 1866):31 (Excelsior Stamp Mill).
49. Krom Rolls are described in S.R. Krom, "Improvements in Ore-Crushing Machinery," AIME *Trans* 14 (1885-86):497-508; and "Report of Rossiter Raymond on the Mechanical Appliances of Mining," pp. 651-55.
50. See Blake's own description of both the original and the improved version of the Blake fine crusher in Theodore A. Blake, "The Blake System of Fine Crushing," AIME *Trans* 13 (1884-85):232-48. See also "Report of Rossiter Raymond on the Mechanical Appliances of Mining," pp. 647-52.
51. Rathbone, "On Copper Mining in the Lake Superior District," pp. 68-123; and *Canadian Patent Office Record* 14 (1888).
52. See Young, *Western Mining,* pp. 66-75.
53. See, for example, *Scientific American* 7, 46 (31 July 1852):364 (Cochran's Quartz Crushing Machine); 9, 15 (24 December 1853):120 (Collyer's California Quartz Crusher); and 9, 20 (14 November 1865):305 (Improved Chilean-Type Quartz Mill).
54. The Wheeler amalgamating pan was the oldest technique in use in American western mining. See *Engineering* 27 (7 March 1879):190-91. For a description of the popular Washoe process, see A.D. Hodges, Jr., "Amalgamation at the Comstock Lode, Nevada: A Historical Sketch of Milling Operations at Washoe, and an Account of the Treatment of Tailings at the Lyon Mill, Dayton," AIME *Trans* 19 (1890-91):195-231; and also "Report of Rossiter Raymond on the Mechanical Appliances of Mining," pp. 683-91.
55. The best-known pulverizer was the Dingley, developed in Cornwall sometime before the year 1873. See Henry T. Ferguson, "On the Mechanical Appliances Used for Dressing Tin and Copper Ores in Cornwall,"

IME *Proc* pt. 1 (July 1873):119-52; and *The School of Mines Quarterly* 4, 4 (1883):306-7. T.B. Jordon Sons & Commans (an English firm) manufactured and marketed their own patented pulverizer (Canada Patent No. 18,928, 20 March 1884), *Engineering* 31 (10 December 1884):542 (advertisement).

56. The Carr Disintegrator was developed in Bristol, England (1871) for granulating superphosphate of lime. It was also used in flour-milling and for granulating clay. See *Scientific American* 37 (4 August 1877):67.
57. See Taggart, "Seventy-Five Years of Progress in Ore-Dressing," pp. 95-101. The work of the magnets was usually aided by air currents or a stream of water. See, for example, Ransom Cooke's Magnetic Separator, *Scientific American* 3 (17 June 1848):305, 308; Edison's Unipolar Non-Contact Electric Separator, AIME *Trans* 17 (1888-89):741-44; and the Ball-Norton Electro-Magnetic Separator, AIME *Trans* 19 (1890-91):192-93.
58. A thorough historical account of the evolution from hand sieves to intermittent, non-continuous, self-acting, hydraulic jiggers and of their spread throughout the mining district of the world is contained in Philip Argall, "Continuous Jigging Machinery," MIC *Proc,* 1, 9 (1884):337-51. The wider range of Cornish ore-dressing techniques, such as the Borlase Buddle and percussion tables, are detailed in Frenchville, "The Results Obtained by the Cornish System of Dressing Tin Ores," pp. 93-104; Rouse, "Borlase's Cornwall," pp. 879-83; and "Report of Rossiter Raymond on the Mechanical Appliances of Mining," pp. 701-10. The application of traditional Cornish ore-dressing techniques in the early days of Michigan mining is described in William Stevens, "The Prospects of the Lake Superior Mining Region," *Mining Magazine* 2, 2 (1854):149-53.
59. Collom's Jigging Machinery was developed in the Michigan mining region by a Cornishman, John Collom, in the 1850s. For more details, see Argall, "Continuous Jigging Machinery," pp. 351-55; Charles M. Rolker, "The Allouez Mine and Ore Dressing as is Practiced in the Lake Superior Copper District," AIME *Trans* 5 (1876-77):593; Rathbone, "On Copper Mining in the Lake Superior District," pp. 102-3. Ferguson, "On the Mechanical Appliances Used for Dressing Tin and Copper Ores in Cornwall," p. 135, carries a diagram of the Collom Washer. The Edward Washer was developed at the Portage Mine by its captain. (See *Mining Magazine* [August 1857]:184-85; and Engineering 4 [20 November 1857]:379). Later, improved versions included Krom's Air-Jig, and its improved version, the Paddock Air-Jig, *Engineering and Mining Journal* 42 (August 1886):7-8. Very popular was the Hartz Fininshing Jig developed in the late 1870s; see NEIME *Trans* 29 (1879-80):159-99.
60. The Borlase Buddle is described in Ferguson, "On the Mechanical Appliances Used for Dressing Tin and Copper Ores in Cornwall," pp. 125-26. Its use and improvement at the Ontonagon Mine in Michigan is described in Rathbone, "On Copper Mining in the Lake Superior District," pp. 89-96, n. 61. Not all mining locations had an adequate supply of water with which to operate on the wet process; therefore some dry separation techniques were experimented with in North America. See, for example, S.J. Pearce's Centrifugal Ore Concentrator, which was patented simultaneously in Buckingham, Quebec, and in New York City (Canada Patent No. 2,719, 4 August 1868), *Scientific American* 20 (27 February 1869): 129.
61. Experiments in the 1880s with a series of washers, jigs, and bumping tables at the St. Joseph Lead Works in Missouri were well-publicized in their day. See H.S. Monroe, "The New Dressing Works of the St. Joseph Lead Company, at Bonne Terre, Missouri," *Canadian Mining Review* 9, 9 (1890):127-30. One popular innovation was the Parsons-Rittinger Table, a modified side-bump table (C.B. Parsons was the superintendent of the St. Joseph Lead Works). Improvements included the Golden Gate Concentrator (Canada Patent No. 20,458, 3 November 1873), *The School of Mines Quarterly* 8, 4 (1887):351-59; and the Krause Separation Table (developed in the Michigan mines), *Canadian Mining Review* 9, 7 (July 1890):102.
62. See *Mechanic's Magazine* 36, 981 (28 May 1842):447, and 43, 1146 (1845):57. For a description and diagram, see *Scientific American* 2 (5 June 1847):289. Bruton was a civil engineer.
63. For details see Chapter 4, and Dianne Newell, "'All in a Day's Work': Local Invention on the Ontario Mining Frontier," *Technology and Culture* (forthcoming). Brief mention is made in Young, *Western Mining,* p. 139. Frue's association with the Silver Islet Mine is not generally recognized in the secondary liter-

ature on mining technology. Canada Patent No. 3,974, 26 October 1874.

64. See *Ontario Commission on Mineral Resources*, p. 375ff. Edward D. Peters, Jr., *Modern American Methods of Copper Smelting* (New York: Scientific Publishing, 1887), pp. 37-81, is one of the best reference works on developments in historical metallurgy to that decade. See also Frederick Laist, "Seventy-Five Years of Progress in Smelting and Leaching," in *Seventy-Five Years of Progress in the Mineral Industry*, pp. 131-32.
65. Peters, *Modern American Methods of Copper Smelting*, pp. 109-114.
66. One of the best-known revolving furnaces was the Brückner. (See ibid., p. 128; "Report of Rossiter Raymond on Metallurgical Processes," pp. 748-55; and J.M. Locke, "The Brückner Revolving Furnace," AIME *Trans* 2 [1874]:295-99).
67. *Ontario Commission on Mineral Resources*, pp. 274-76; and Peters, *Modern American Methods of Copper Smelting*, pp. 293-96; and Laist, "Smelting," pp. 133-35.
68. Peters, *Modern American Methods of Copper Smelting*, pp. 215-77; and W.M. Courtis, "The Wyandotte Silver Smelting and Refining Works," AIME *Trans* 2 (1873-74):89-100.
69. See Fraser and Chalmers Company, *Catalogue, No. 3* (Chicago, c. 1888); E.D. Peters, Jr., "The Sudbury Ore Deposits," AIME *Trans* 18 (1889-90):278-89; and E.D. Peters, Jr., "Sudbury Mines and Works," *Canadian Mining Review* 8, 10 (1889):135.
70. See description *Ontario Commission on Mineral Resources*, pp. 145-46.
71. Perhaps the most useful study is provided by Jeanne McHugh, *Alexander Holley and the Makers of Steel* (Baltimore: Johns Hopkins University Press, 1980). See also Charles K. Hyde, *Technological Change and the British Iron Industry, 1700-1870* (Princeton: Princeton University Press, 1977), for a more specialized account of early developments; and a rather lengthy summary of developments to the 1880s in *Ontario Commission on Mineral Resources*, pp. 319-69.
72. A.W.G. Wilson, *Development of Chemical, Metallurgical and Allied Industries in Canada* (Ottawa: Department of Mines, 1924), cited in William M. Marr and Donald G. Paterson, *Canada: An Economic History* (Toronto: Macmillan, 1980), pp. 356, 366-67; and W.J.A. Donald, *The Canadian Iron and Steel Industry: A Study in the Economic History of a Protected Industry* (Boston: Houghton Mifflin, 1915).
73. Described in Young, *Western Mining*, pp. 199, 273-75; and "Report of Rossiter Raymond on Metallurgical Processes," pp. 730-33; and see also Small, "Notes on the Stamp Mill and Chlorination Works of the Plymouth Consolidated Gold Mine," pp. 305-8. The role of Australian mining operations in developing chlorination techniques is detailed in G. Blainey, *The Rush that Never Ended: A History of Australian Mining* (Melbourne: Melbourne University Press, 1963).
74. Young, *Western Mining*, pp. 274-75.
75. A non-technical, highly readable account of this is to be found in D.L. Bumstead, "Copper Smelting in Canada," *CIM Bulletin* 75, 846 (1982):36-40.
76. A thorough, if somewhat disorganized, history has been written by Brantly, *History of Oil Well Drilling* (See, for example, pp. 489-90). For an overview of the changing techniques of production in the American petroleum industry see Harold F. Williamson and Arnold P. Daum, *The American Petroleum Industry*, vol. 1, *The Age of Illumination 1859-1899* (Evanston: Northwestern University Press, 1959).
77. Brantly, *History of Oil Well Drilling*, p. 59.
78. See, for example, the Beart Boring Apparatus, developed in England in 1844, that employed a stream of water to remove debris (*Mechanics' Magazine* 42, 1123 [1845]:111); and Gard's Patent Boring and Sinking Machinery, a Cornish percussion borer also developed in England, in 1846 (*Mechanics' Magazine* 48, 1291 [1848]:433-35; and *Scientific American* 5, 18 (1850):137.

    The Kind and de Wende's Improvements in the Process and Instruments for Boring Earth and Sinking Shafts, patented in United Kingdom, in 1851. S.H. Blackwell, "On Kind's Improved System of Boring," IME *Proc* (1854):87; and *Mechanics' Magazine* 55, 1461 (1851):119. The Kind-Chaudron Process for Sinking and Tubing Mine Shafts, patented in UK, in 1857, was employed in collieries in Belgium and Germany, and in brine wells, in Staffordshire. Julien Degy, "The Kind-Chaudron Process for Sinking and Tubing Mine Shafts," AIME *Trans* 5 (1876-77):117-31; and NEIME *Trans* 2 (1853-54):60-62.
79. See an excellent account of this development by E.D. Thoenon in, "Petroleum Industry in West Virginia before 1900" (Ph.D. dissertation, University of West Virginia, 1956), chapter 4.

80. Ibid., pp. 98-99; Brantly, *History of Oil Well Drilling,* introduction; and N.R. Ball, "Canadian Petroleum Technology during the 1860's" (Master's thesis, University of Toronto, 1972), pp. 81-89ff.
81. The most comprehensive source on this topic is Brantly, *History of Oil Well Drilling.* The "American System" was spread to Baku by Pennsylvania drillers in 1875 ("Petroleum at Baku," *Scientific American* 40, 11 [1884]:166; 40, 22 [1884]:342-43).
82. *Ontario Commission on Mineral Resources,* p. 167 (Duncan Sinclair).
83. Ibid. For details on MacGarvey, see Chapter 6.
84. See *Ontario Commission on Mineral Resources,* pp. 154, 157; Robert Bell, "The Petroleum Fields of Ontario," *Proceedings and Transactions of the Royal Society of Canada* 5 (1888):111; and Dianne Newell, "Technological Innovation and Persistence in the Ontario Oilfields: Some Evidence from Industrial Archaeology," *World Archaeology* 15, 2 (1983): 184-95.
85. The Stangenkunst is detailed in Robert P. Multauf, "Mine Pumping in Agricola's Time and Later," Paper 7, USM Bulletin 218, *Contributions from the Museum of History and Technology* (1959):114-20, and its adaptation in Ontario petroleum is described in Newell, "Technological Innovation and Persistence," pp. 186-93.
86. See A.M. Johnson, "The Development of American Petroleum Pipe Lines: A Study in Enterprise and Public Policy, 1862-1906" (Ph.D. dissertation, Vanderbilt University, 1954). As the title suggests, the focus is more on the politics of pipelines than on the technological aspects. See also *Ontario Commission on Mineral Resources,* pp. 163-66, and Everette L. DeGolyer, "Seventy-Five Years of Progress in Petroleum" in *Seventy-Five Years of Progress in the Mineral Industry,* pp. 295-99 (contains numerous errors).
87. B. Martell, "On the Carriage of Petroleum in Bulk on Over-Sea Voyages," *Engineering* 42 (30 July 1887):107-11.
88. *Ontario Commission on Mineral Resources,* p. 183. For an historical overview of Ontario salt extraction see D.F. Hewitt, "Salt in Ontario," *Industrial Minerals Report No. 6* (Toronto: Department of Mines, 1962).
89. See Johnson, "Expansion of the Petroleum and Chemical Industries, p. 666; and the testimonials of James Kerr, a Petrolia oil-refiner with connections in the oil business dating back to 1862, and Martin J. Woodward, in *Ontario Commission on Mineral Resources,* pp. 163, 157.
90. Williamson and Daum, *The American Petroleum Industry,* p. 282; and Ida M. Tarabell, *The History of Standard Oil Company* (New York: Norton, 1969), p. 11. Both are cited in Sally F. Zerker, "Centralization in the International Oil Industry" (paper read before the Annual Meeting of the Canadian Historical Association, Vancouver, 6-8 June 1983), p. 18.
91. The problem is outlined in *Ontario Commission on Mineral Resources,* pp. 162-65; and the innovations discussed in Newell, "'All in a Day's Work': Local Invention on the Ontario Mining Frontier." For details on Frasch, see Chapter 6.
92. See T.S. Hunt, "On the Goderich Salt Region . . . ," in GSC *Report of Progress, 1866-69* (1870), pp. 212-42; J.L. Smith, "Observations on the History of the Trade and Manufacture of Canadian Salt," in GSC *Report of Progress, 1874-75* (1876), pp. 294-95; and *Engineering* (8 August 1879):103-4.
93. See Jennifer Tann, "Fuel-Saving in the Process Industries during the Industrial Revolution: A Study in Technological Diffusion," *Business History* 15 (1973):147-59.
94. These innovations are discussed in Chapter 6 and summarized in Newell, "'All in a Day's Work': Local Invention on the Ontario Mining Frontier."

NOTES TO CHAPTER THREE

1. See discussion of the diffusion of new techniques in William M. Marr and Donald G. Paterson, *Canada: An Economic History* (Toronto: Macmillan 1980), pp. 29-40. Examples of the importation of embodied technique in Canada are detailed in Elwood S. Moore, *American Influence in Canadian Mining* (Toronto: University of Toronto Press, 1941); and Richard Pomfret, "Mechanization of Reaping in Nineteenth Century Ontario: A Case Study of the Pace and Causes of the Diffusion of Embodied Technical Change," *Journal of Economic History* 36 (June 1976):399-415. The influence of foreign capital on technique choice is dealt with throughout Clark C. Spence's monograph, *British Investments and the American Mining Frontier, 1860-1891* (Ithaca: Cornell University Press, 1958).

2. On mining engineering see J.C. Gwillum, "The Status of the Mining Profession," *Journal of the Canadian Mining Institute* 10 (1907): 321-39; William B. Potter, "Some Thoughts Relating to the American Institute of Mining Engineers and its Mission," (Presidential Address), *Transactions of the American Institute of Mining Engineers* [AIME *Trans*] 17 (1888-89):485-94; Clark C. Spence, *Mining Engineers and the American West: The Lace-Boot Brigade, 1849-1933* (New Haven: Yale University Press, 1970), chapters 1 and 2; and John B. Rae, "The Invention of Invention," in *Technology in Western Civilization,* vol. 1, eds. Melvin Kransberg and Carroll W. Pursell Jr. (New York: Oxford University Press, 1967), pp. 327-36. For a recent account of the science of geology in the nineteenth century, see Morris Zaslow, *Reading the Rocks: The Story of the Geological Survey of Canada, 1842-1972* (Toronto: Macmillan, 1975), especially pp. 1-34.

   Little has been published on engineering education in Canada. See the discussion on engineers and rational technology in Bruce Sinclair et al., eds., *Let Us Be Honest and Modest. Technology and Society in Canadian History* (Toronto: Oxford University Press, 1974), pp. 221-23; C.R. Young, *Early Engineering Education at Toronto, 1851-1919* (Toronto: University of Toronto Press, 1958); Donald MacLeod, "Mines, Mining Men, and Mining Reformers: Changing the Technology of Nova Scotian Gold Mines and Collieries, 1858 to 1910". (Ph.D. dissertation, University of Toronto, 1981), especially chapters 9 and 10; Dianne Newell, "Professionalization of Mining Engineers in Late Nineteenth-Century Canada" (paper read before the Workshop on Professionalization in Modern Society, University of Western Ontario, 13-15 March 1981); and Louis Henry, "The Training of a Mining Engineer," *Canadian Mining Review* 14, 2 (1895): 220-21.
3. See "Essay on Sources" in Zaslow, *Reading the Rocks,* pp. 541-44; and Dianne Newell, "Published Government Documents as a Source for Interdisciplinary History: A Canadian Case Study," *Government Publications Review* 8A (1981):381-93.
4. See, for example, Henry C. Bolton, *A Catalogue of Scientific and Technical Periodicals (1665-1882)* (Washington: Smithsonian Institution, 1885); and National Research Council of Canada, Canada Institute for Scientific and Technical Information, *Union List of Scientific Serials in Canadian Libraries,* 2 vols. 6th ed. (Ottawa: Queen's Printer, 1975).
5. See Eugene S. Ferguson, "Expositions of Technology, 1851-1900," in *Technology in Western Civilization,* vol. 1, pp. 706-25.
6. Dianne Newell, "Canada at the World's Fairs, 1851-1876," *Canadian Collector* 11, 4 (1976):13.
7. See Spence, *Mining Engineers;* and Carroll W. Pursell, Jr., *Early Stationary Steam Engines in America: A Study in the Migration of a Technology* (Washington: Smithsonian Institution Press, 1969). A number of Canadians belonged to the AIME (formed 1872), and the meetings occasionally were held in Canada (Montreal, 1879; Halifax, 1885; and Ottawa, 1889). See "List of the Meetings . . . to June 1890," AIME *Trans* 18 (1889-90), and various issues for published lists of members.
8. Moore, *American Influence,* p. 98.
9. Standard literature on the subject of Cornish migrations includes John Rowe, *Cornwall in the Age of the Industrial Revolution* (Liverpool: University Press, 1972); A.K. Rouse, "Borlase's Cornwall," *History Today* 23, 12 (1973):879-83; and John Rowe, *The Hard Rock Men: Cornish Immigrants and the North American Mining Frontier* (Liverpool: University Press, 1974). The latter contains almost no reference to Canadian mining districts. Cornishmen were important, too, in the Australian fields. See G. Blainey, *The Rush that Never Ended: A History of Australian Mining* (Melbourne: Melbourne University Press, 1963), chapter 11. Blainey claims that the copper-mining towns of Moonta and Dakina were possibly the largest Cornish communities in the nineteenth century outside Great Britain (p. 119).
10. Macdonell and Foster Papers, MG III 40, Letterbook 1870-1871, PAC. Macdonell to the Chief Administrator of the Nova Scotia Mines, 4 February 1871.
11. Ibid., Fred Foster to Dean and Williams, 29 December 1870.
12. Ibid., Fred Foster to Macdonell, 3 February 1871.
13. Ibid., Fred Foster to Dean and Williams, 29 December 1870.
14. Ibid., Fred Foster to Macdonell, 3 February 1871. He suggested that copies be sent to the same mining men who had been sent promotional literature on the battery.
15. Ibid., Fred Foster to I.W. Forbes, 2 February 1871.

16. Ibid., Fred Foster to Charles Foster, 15 April 1871.
17. Ibid.; and Macdonell to C.A. Trowbridge, 25 June 1871.
18. Bernard Drell, "The Role of the Goodman Manufacturing Company in the Mechanization of Coal Mining," (Ph.D. dissertation, University of Chicago, 1930), pp. 18-19, 22, 26, 30.
19. Great Britain, Royal Commission on Technical Instruction, *Report on Technical Education in the United States of America and Canada,* by William Mather (1881). Reprinted in *Irish University Press Series of British Parliamentary Papers* (Shannon: Irish University Press, 1970), vol. 6. See also, Ontario, Royal Commission on Mineral Resources, *Report* (Toronto: Warwick and Sons, 1890) [hereafter *Ontario Commission on Mineral Resources*], pp. 491-509; Spence, in *Mining Engineers,* p. 25 and chapter 2, discusses the evolution of early mining schools; as does Charles Twite, in "Technical Education of Miners," *Proceedings of the Mining Institute of Cornwall* MIC *Proc* 1, 7 (1883):201-14. Twite claimed that mining was the first industry to provide technical education (beginning in 1851 with the London School of Mines).
20. For the Canadian case, see *Annual Report of the Geological Survey of Canada* [hereafter GSC, *Annual Report*], 1870, p. 12; and Zaslow, *Reading the Rocks,* p. 246. For the United States, see U.S., 40th Congress, 3d Session, 1869, *Report of Rossiter W. Raymond on the Mineral Resources of the States and Territories West of the Rocky Mountains,* House Exec. Doc. no. 54, sect. 7 (Mining Education), pp. 224-33.
21. See Samuel B. Christy, "The Growth of American Mining Schools and Their Relation to the Mining Industry," AIME *Trans* 23 (1893):445; John A. Church, *Mining Schools in the United States* (New York: Waldron and Payne, 1871); James McGivern, *First Hundred Years of Engineering Education in the United States* (Washington: Gonzaga University Press, 1960); and R.H. Richards, "American Mining Schools," AIME *Trans* 15 (1886-87): 309-40, 809-19. See also, *Ontario Commission on Mineral Resources,* pp. 509-12. Good comparative data is to be found in M.E. Wadsworth, "Some Statistics on Engineering Education," AIME *Trans* 27 (1897):712-31; and Great Britain, *Report on Technical Education in the United States of America and Canada.* The Columbia School of Mines, for example, lists only one graduate before 1889 with a Canadian address. See "Regular and Associate Members and All Other Graduates," *The School of Mines Quarterly* 9, 4 (1889):vi.
22. The school enjoyed the benefit of the Geological Survey's Montreal location until it moved to Ottawa in 1881. See Spence, *Mining Engineers,* p. 34; and McGill University, *Graduates of McGill University, Montreal, Corrected to December 1900* (Montreal: Witness Printing, 1901). Seventy-seven students graduated as mining engineers by 1900, but only seventeen of them before 1890 (five in the 1870s, twelve in the 1880s, and sixty in the 1890s). See Newell, "Professionalization of Mining Engineers in the Late Nineteenth Century," Table 1.
23. Ontario, *Statutes,* 36 Vict., 1873, c. 30; and C.R. Young, *Early Engineering Education at Toronto, 1851-1919* (Toronto: University of Toronto Press, 1958), pp. 97, 106.
24. Legislative Assembly of Ontario, *Sessional Papers,* 1878, "Report of the School of Practical Science for 1877."
25. See Note 22.
26. Ibid. The mobility of these graduates is remarkable. In cases where mining engineers were members of the Canadian Institute of Civil Engineers (the only professional engineering association in Canada during the period under study), their careers have been recorded and preserved in the Engineering Institute of Canada Collection, Membership Files, PAC.
27. Morris Zaslow, "Edward Barnes Borron, 1820-1915, Northern Pioneer and Public Servant Extraordinary," in *Aspects of Nineteenth-Century Ontario,* ed. Frederick H. Armstrong, et al. (Toronto: University of Toronto Press, 1974), pp. 297-311.
28. Canada (Province), "Report of the Commissioner of Crown Lands for 1865," (1866), Appendix No. 24.
29. Zaslow, "Edward Barnes Borron," p. 311, for a full list of Borron's publications.
30. Borron testified during his last year as stipendiary magistrate. See *Ontario Commission on Mineral Resources,* pp. 69, 92, 114, 147, 194, 238, 316, 402, and 428.
31. Zaslow, *Reading the Rocks,* pp. 560, 138.
32. W.M. Courtis, "The Wyandotte Silver Smelting and Refining Works," AIME *Trans* 2 (1873-74):89-100.
33. "Macfarlane's New and Useful Improvement on the Art of Extracting Copper from Its

Ores by the Humid Process," 4 December 1868, Canada Patent No. 2889. A collection of Macfarlane's papers is housed in the rare book collection at McGill University.

34. See Young, *Early Engineering Education,* pp. 29-33, 73-75ff; and Zaslow, *Reading the Rocks,* p. 78.
35. "Chapman's Art of Treating Auriferous Mispickel for the Extraction of Gold, and Art of Producing Paint during the Treatment of Auriferous Mispickel for the Extraction of Gold," Canada Patent Nos. 2,015, 29 January 1873; and 2,026, 3 February 1873.
36. H.J. Morgan, ed., *The Canadian Men and Women of Their Time: A Handbook of Canadian Biography of Living Characters,* 2nd ed. (Toronto: Wm. Briggs, 1912), p. 799.
37. See Edward Peters, Jr., *Modern American Methods of Copper Smelting* (New York: Scientific Publishing, 1887).
38. Zaslow, *Reading the Rocks,* p. 38, and chapter 3. A collection of Logan's papers is housed in the McGill University Archives.
39. Professions and periods of service for Survey staff are taken from Zaslow, *Reading the rocks,* appendix 2, pp. 555-64. The publications authored by them are listed in A.G. Johnson, comp., *Index of Publications, Geological Survey of Canada (1845-1958)* (Ottawa: Department of Mines and Technical Surveys, 1961), pp. 1-19.
40. Hunt performed analyses for the Survey prior to his appointment and also published extensively in numerous professional journals of the day.
41. Robert Bell Papers, MG 29, B 15, file 29 (bibliography), PAC.
42. Zaslow, *Reading the Rocks,* pp. 142, 132, and 557; and Canada, Department of Mines, *Reports on Natural Gas and Petroleum in Ontario Prior to 1891* by H.P.H. Brumel (Ottawa: Queen's Printer, 1892). Coste was interested in the Medicine Hat, Alberta, region.
43. In his first years he undertook a number of major mining monographs, one of the most important of which was a combined historical study and status report on silver mining in the Lake Superior region. See also Ingall's 1888 proposal for membership in the Engineering institute of Canada Collection, MG 28 I 277, vol. 7. PAC.
44. See, for example, S.H. Scudder, *Catalogue of Scientific Serials of all Countries including the Transactions of Learned Societies in Natural, Physical, and Mathematical Sciences, 1633 to 1876* (Cambridge, 1879); and Bolton, *A Catalogue of Scientific and Technical Periodicals.* For one useful guide to contemporary mining monographs, see Professor H.S. Monroe, "List of Books on Mining," *School of Mines Quarterly* 10, 2 (1889):176-84. An excellent indication of the literature on mining available to Canadian professionals is found in the annual lists of additions to the library, published in the Geological Survey, *Annual Reports,* beginning with the 1873 report.
45. The reference for this selection is National Research Council of Canada, *Union List of Scientific Serials.*
46. The title was changed to *Canadian Mining Journal* in 1907.
47. See Peter J. Bowler, "The Early Development of Scientific Societies in Canada," in *The Pursuit of Knowledge in the Early American Republic: American Learned Societies from Colonial Times to the Civil War,* ed. A. Oleson and S.C. Brown (Baltimore: Johns Hopkins Press, 1976), pp. 326-29; Trevor H. Levere and Richard A. Jarrell, eds., *A Curious Field-Book: Science and Society in Canadian History* (Toronto: Oxford University Press, 1974), introduction; and Vittorio M. G. DeVecchi, "Science and Government in Nineteenth-Century Canada" (Ph.D. dissertation, University of Toronto, 1978).
48. The Survey was formed in 1841 as a temporary creation and was kept aloft by a series of acts. For reports, see Canada (Province), Geological Survey, *Report of Progress, 1844-1863* (1846-1866). See also O.B. Bishop, *Publications of the Government of the Province of Canada, 1841-1867* (Ottawa: National Library, 1963), pp. 295-99; and Johnson, *Index of Publications, Geological Survey of Canada,* pp. 1-18.
49. "Catalogue of Metals and their Ores, etc., with Locations, exhibited at the Exhibition of the Industry of All Nations, London," in GSC *Report of Progress 1850-51* (1852), pp. 54-63; and "A Sketch of the Geology of Canada Serving to explain the Geological Map and the Collection of Economic Minerals sent to the Universal Exhibition at Paris, 1855, by W.E. Logan and T. Sterry Hunt," in *Canada at the Universal Exhibition of 1855* (Toronto, 1856).
50. "Report of the Select Committee Appointed to Report upon the Best Means of Making Public the Valuable Information Already Obtained by the Geological Survey, and Completing it at an Early Period upon

an Uniform System," in GSC *Report of Progress 1852-53* (1854).

51. See *Transactions of the Canadian Society of Civil Engineers* 1-2 (1887-88):6-7.
52. "Record of Mines and Mineral Statistics, by Chas. Robb," in GSC, *Report of Progress, 1871-72* (1872), pp. 146-54. See also the explanation given in the introduction to "Statistical Report, 1886, by Eugène Coste," in GSC *Summary Report 1886* (1887); and a similar report on the problems of collecting statistical returns from the mining industry in the United States, in United States, *10th Census,* 1880 (1883), pt. 1, pp. 28-29. For a later day overview, see S.P. Malholtra, "The Development of Canadian Mineral Statistics and Confidentiality under Federal and Provincial Statistical Acts" (Canada, Dept. of Energy, Mines and Resources, Mineral Resources Branch, MR136, 1973).
53. See Canada, House of Commons, *Debates,* 1884, p. 575; and "Report of the Select Committee ... to Obtain information as to the Geological Survey ...," in Canada, House of Commons, *Journals,* 1884, App. No. 8.
54. "Report on Mining and Mineral Statistics of Canada for the Year 1888," in GSC *Annual Report,* n.s., vol. 4, 1888-89, "S" (1889).
55. RG 87, Department of Mines, Mineral Statistics, PAC.

NOTES TO CHAPTER FOUR

1. See Warren J. Jestin, "Provincial Policy and the Development of the Metallic Mining Industry in Northern Ontario: 1845-1920" (Ph.D. dissertation, University of Toronto, 1977), chapter 3; and H.V. Nelles, *The Politics of Development: Forest, Mines, and Hydro-Electric Power in Ontario, 1849-1914* (Toronto: Macmillan, 1975), pp. 19-31.
2. See T.A. Rickard, *The Copper Mines of Lake Superior* (New York: Engineering and Mining Journal, 1905); and W.B. Gates, Jr., *Michigan Copper and Boston Dollars: An Economic History of the Michigan Copper Industry* (Cambridge, Mass.: Harvard University Press, 1951). Both works discuss the problems in developing the Lake Superior mines caused by the bottleneck at Sault Ste. Marie.
3. Canada (Province), *Report of the Commissioner of Crown Lands, 1865* (1866), App. 24. "Land and Road Surveys for the Lakehead District Commenced in 1865." See also "Return ... Shewing the names of each Colonization Road on which Provincial money has been expended since July 1st, 1867; its length in miles; the constituency ... in which it is situated; the amount expended on it each year, distinguishing between amounts for repairs and amounts of new road ...," in Ontario, *Sessional Papers,* 1885, No. 24.
4. See Elizabeth D. Arthur, ed., *The Thunder Bay District, 1821-1892. A Collection of Documents* (Toronto: University of Toronto Press, 1973), pp. lxxvii-lxxviii; and Edwin C. Guillet, *Pioneer Travel in Upper Canada.* 1933. Reprint (Toronto: University of Toronto Press, 1963), chapters 7 and 8.
5. G.N. Fuller, *Geological Reports of Douglass Houghton, First State Geologist of Michigan, 1837-1845* (Lansing, Mich., 1941), p. 59. See also the discussion of the implications of Houghton's explorations for the Canadian side, in William Hamilton Merritt Papers, Package No. 32, "Mines and Mining, 1845-1865," George K. Smith to W.H. Merritt, 17 September 1845, PAO [hereafter, W.H. Merritt Papers]; and "On the Geology and Economic Minerals of the North Shore of Lake Superior with Remarks on Copper Ores and their Treatment, by W.E. Logan," in GSC *Report of Progress, 1846-47* (1847) [hereafter, "Logan's Report of Lake superior, 1847"]. The prehistoric Old Copper Indians of the Upper Great Lakes region had been miners and fabricators of pure, or native, copper. Copper was mined from thousands of shallow pits, mostly in the upper peninsula of Michigan and along the Ontario shores of Lake Superior. This activity was mentioned in the narratives of Jesuit missionaries in the seventeenth century, and some mining of native copper on the north shore of Lake Superior was attempted in the 1770s by Europeans. See Ontario, Royal Commission on Mineral Resources, *Report* (Toronto: Warwick and Sons, 1890) [hereafter, *Ontario Commission on Mineral Resources*], pp. 92-93.
6. For the government's correspondence with petitioners, see W.H. Merritt Papers, and British North American Mining Company *Minute Book,* 1846-1910, PAC [hereafter, BNA Mining Co., *Minute Book*]. Lists of individual location holders appeared periodically in government reports. See "Tabular Return of the Persons who have received licenses for opening and working

mines on Lakes Huron and Superior . . . ," in Canada (Province), *Appendix to the Journals*, 1851, No. 2, Appendix W; and "Return . . . shewing the names of all persons to whom either patents or licenses have been granted of mineral . . . lands on the North Shores of Lakes Huron and Superior . . . ," in Canada (Province), *Sessional Papers*, 1860, No. 39. See also "List of the Mining Grants made by the Department of Crown Lands of the Province of Canda, 1852-67" (unpublished manuscript, Lands Survey office, Toronto). The most detailed secondary account occurs in Margaret A. Wright, "The Canadian Frontier, 1840-1867" (Ph.D. dissertation, University of Toronto, 1943), pp. 37-40.

The list of original shareholders and directors of companies formed to work the Lake Huron and Superior mineral locations reads like a "who's who" for the Province of Canada. George Desbarats (Montreal) was a major shareholder (1846) in the Montreal Mining company, a director and trustee (1846, 1847) of the British and Canadian Mining Company, a director and trustee (1847) of the Garden River Copper Mining Company, and a director (1849) of the Huron Mining Company, for example, Likewise, Stewart Derbishire (Montreal) was a shareholder (1846) of the Montreal Mining Company, president (1847) of the Huron and St. Mary's Copper company, and a director (1849) of the Huron Copper Bay Company. Desbarats and Derbishire were the Queen's printers. And John Prince (Sandwich) was chairman and principal shareholder of the British North American Mining Company, which had been formed to work his location on Lake Superior, an original shareholder in the Montreal Mining Company, and a director (1849) of the Huron Copper Bay Company. Other prominent capitalists who were associated with more than a single mining venture in this district include Theodore Hart, Peter McGill, George Moffat, James Ferrier, W.C. Meredith, and the Hon. Francis Hinks, all of Montreal. This information has been gleaned from the lists noted above and the acts of incorporation for mining companies reproduced in Canada (Province), *Statutes*, 1846 to 1850.

7. See W.H. Merritt Papers, Executive Council Committee Report, 19 September 1845; 27 October 1845; 12 November 1845; and 10 December 1845. See also various correspondence between W.H. Merritt and the Province of Canada government over the license application for George Smith, a resident of Michigan. This restriction was not applied in all cases, however. See list of shareholders of the BNA Mining Co. (*Minute Book*).
8. See W.H. Merritt Papers. Correspondence between George Smith and W.H. Merritt, and from Trowbridge to Merritt, 11 December 1845, in which Trowbridge offers to serve as United States banker for the company and forwards samples of United States mining company organization and stock certificate schemes. See also Trowbridge to Merritt, 4 and 7 January 1846, on the flooded state of the United States copper market; E. Ryan to Merritt, 24 January 1846, on the "gloom" over the English money market; George Smith to Merritt, 4 June 1846, on the poor money markets in Boston and New York; and 11 December 1846, on the failure to sell Montreal Mining Co. stock in Windsor, CW, at "any price owing to the scarcity of money."
9. The definitive study on the early history of provincial mining policy is provided by W.J. Jestin, "Provincial Policy."
10. See Note 6.
11. BNA Mining Co., *Minute Book*. The company established itself in Montreal, 30 April 1846. The directors of the company were British subjects, but the majority of stockholders lived in the United States (principally Michigan). It is not known whether or not the company concealed this fact from the legislature. See also British North American Mining Co., *Report of the Directors of the British North American Mining Company* (Montreal: Lovell and Gibson, 1847, 1848).
12. BNA Mining Co., *Minute Book*, 14 December 1847 (report of Mr. Robertson).
13. See report on the Quebec and Lake Superior Mining Company (incorporated 28 July 1847), in Allan Macdonell Mining Papers, MG 24, 1-8, PAC. By the mid-1860s most of their holdings had been forfeited. The bulk of the company shares eventually fell into the hands of a few parties in Quebec. Their Michipicoten and St. Ignace Islands locations were taken over in 1881 by English capitalists (E.D. Arthur, ed., *The Thunder Bay District*, pp. 50-52, 107 [fns., pp. 50, 56, 107]).
14. See Walter William Palmer, "A Pioneer's

Mining Experience on Lake Superior and Lake Huron," in Ontario Bureau of Mines, *Report, 1892* (1893), pp. 171-78. Palmer was mining captain for these operations. He planned to introduce a Chilean system of copper smelting, but the sites were abandoned because of the poorness of the deposits and the "hostility" of the local "half-breeds."

15. Ibid., pp. 175-76; and Hudson's Bay Company Papers, Outward Correspondence, Sault Ste. Marie Post, 16 June, 3 September, 12 and 16 November and 15 December 1849; 13 and 17 January and 14 February 1850, Hudson's Bay Company Archives, PAM.
16. See the HBC correspondence cited in Note 15, plus *Ontario Commission on Mineral Resources*, p. 98.
17. Ibid.
18. Ibid., pp. 23, 101; and Montreal Mining Company, *Annual Report of the Directors of the Montreal Mining Company to the Stockholders* [hereafter, *Report of the Montreal Mining Company*] (Montreal, 1846).
19. See Upper Canada Mining Company, *Report of the Upper Canada Mining Company* (Hamilton: The Spectator Office, 1847, 1848). A dozen men (Cornish and German) were hired from abroad. A smelting operation for the mines was considered – it would be on either the Cornish system or the new furnace developed by A.E. Osborne of New York. See also *Ontario Commission on Mineral Resources*, pp. 93-94.
20. Good secondary accounts include H.J. Carnegie Williams, "The Bruce Mines, Ontario, 1846-1906," *Journal of the Canadian Mining Institute* 10 (1907):149-69. The author was manager of the Bruce Mines in 1906. And a review of official reports concerning Bruce Mines is contained in Ontario, *Royal Ontario Nickel Commission* (Toronto: King's Printer, 1917) [hereafter, *Ontario Commission on Nickel*], pp. 20-25.
21. *Ontario Commission on Mineral Resources*, pp. 93-94.
22. *Report of the Montreal Mining Company*, 1848.
23. H.J.C. Williams, "Bruce Mines," pp. 149-50; and W. Palmer, "Pioneer's Mining Experience," p. 175.
24. A good discussion of this is contained in D.L. Bumstead, "Copper Smelting in Canada," *CIM Bulletin* 75, 846 (1982):36-40.
25. *Report of the Montreal Mining Company*, 1848.
26. Ibid., 1848-1850.
27. HBC Correspondence, SSM Post, 16 June, 12 and 16 November 1849, 15 December 1849, 13 and 17 January 1850; and 14 February 1850.
28. *Ontario Commission on Mineral Resources*, p. 94; and *Report of the Montreal Mining Company*, 1848-50.
29. Ontario Bureau of Mines, *Report, 1893* (1894), p. 36.
30. Ibid.
31. See Borron's testimony, *Ontario Commission on Mineral Resources*, p. 30.
32. Ibid., p. 403.
33. "Report made to the Crown Lands Department by Albert Pellew Salter . . . upon the country bordering upon the North Shore of Lake Huron . . . ," in Canada (Province), *Appendix to the Journals of the Legislative Assembly*, 5, 1856, No. 34; and "Return . . . for a copy of the Report of Count de Rottermund [sic] . . . ," in Ibid., No. 37.
34. *Ontario Commission on Mineral Resources*, pp. 94-95; and *Report of the Montreal Mining Company*, 1853-54.
35. For a description of the new jigging system, see "Report of Count de Rottermund." The Lake Superior "system" is described in Charles M. Rolker, "The Allouez Mine and Ore Dressing as is Practiced in the Lake Superior Copper District," *Transactions of the American Institute of Mining Engineers* [AIME *Trans*] 5 (1876-77):584-611; T. Egleson, "Copper Mining on Lake Superior," AIME *Trans* 6 (1877-78):275-312; and Edgar P. Rathbone, "On Copper Mining in the Lake Superior District," *Proceedings of the Institute of Mechanical Engineers* [IME *Proc*] (February, 1887):86-123.
36. *Report of the Montreal Mining Company*, 1853; and John Rowe, *The Hard Rock Men: Cornish Immigrants and the North American Mining Frontier* (Liverpool: The University Press, 1974), p. 175. Rowe notes that a number of Cornishmen had come from Bruce Mines. He cites the bibliographical sketches collected by persons at the Northern State Normal School in 1925-27. These are preserved in the collections of the Marquette Historical Society.
37. *Ontario Commission on Mineral Resources*, p. 95; and *Report of the Montreal Mining Company*, 1853-54.
38. *Report of the Montreal Mining Company*, 1853-54.
39. *Ontario Commission on Mineral Resources*, p. 95.

40. *Report of the Montreal Mining Company,* 1853-54.
41. "The Marquette Iron Region," *School of Mines Quarterly* 3, 3 (1882):197-207; 3, 4 (1882):244-53. This was a fairly universal phenomenon. See Homer Aschman, "The Natural History of a Mine," *Economic Geography* 46 (1970):172-89.
42. See "Report of Inspection of Mining Locations on Lakes Huron and Superior, by Wm. Gibbard," in Canada (Province), "Report of the Commissioner of Crown Lands, 1860," App. No. 29)[hereafter, "Wm. Gibbard's Report, 1862]; and *Report of the Montreal Mining Company,* 1856.
43. Detailed in Borron's testimony, *Ontario Commission on Mineral Resources,* pp. 94-6.
44. Gibbard was commissioned to visit all mining locations, report on whether or not they were being worked, and recommend their forfeiture for non-use or non-payment (see "Wm. Gibbard's Report, 1860").
45. Canada (Province), *Sessional Papers,* 1864, No. 52, "Statement of Mining Locations on North Shores of Lakes Huron and Superior, forfeited or liable to be so . . . and Sales of New Portions of the same."
46. See Note 13.
47. See Crown Lands Department, RG 1, G-1, vol. 7, Correspondence and Memoranda Regarding "Wallace Mine" with Upper Canada Mining Company, 1864-66, PAO.
48. W.B. Gates, Jr., *Michigan Copper and Boston Dollars,* p. 97.
49. "Wm. Gibbard's Report, 1860"; and GSC *Report of Progress from Commencement to 1863* (1863), p. 695.
50. *Ontario Commission on Mineral Resources,* p. 404.
51. Ibid., p. 97; and "William Plummer," *Canadian Mining Review* 9, 6 (1890):82-83.
52. Ontario Bureau of Mines, *Report, 1893,* pp. 136-37.
53. *Ontario Commission on Mineral Resources,* p. 97.
54. D.L. Bumstead, "Copper Smelting in Canada," p. 36; and *Ontario Commission on Mineral Resources,* p. 404.
55. See W.H. Plummer's testimony, *Ontario Commission on Mineral Resources,* p. 404.
56. Ibid., pp. 23, 101; and W.B. Gates, Jr., *Michigan Copper and Boston Dollars,* p. 45. See also Robert Bell Papers, vol. 39, Mining Industry file, "Table Showing the Amount of Copper ores and Copper Sent from the Bruce, Wellington, and Huron Copper Bay Location from 1847 to 1875" PAC.
57. See Note 44; and W.J. Jestin, "Provincial Policy," table 14, pp. 56-57. See also Ontario, *Statutes,* 31 Vict. c. 19, "An Act Respecting Gold and Silver Mines," (repealed 1869).
58. Good detail on the people involved is provided by Marion Henderson, *The McKellar Story* (Thunder Bay, 1981); Elizabeth Arthur, "The Founding Father [Peter McKellar]," *Thunder Bay Historical Museum Society Papers and Records* [TBHMS *Papers and Records*] 11 (1983):10-22; Keith Denis, "Oliver Daunais, The 'Silver King,'" TBHMS *Papers and Records* 2 (1974):12-21; and Thomas Keefer Papers, MG 29, B-10, sheets 54-65 (speech prepared for delivery to Port Arthur citizens in 1885, by mining promoter and lawyer, T.A. Keefer), PAC. See Canada (Province), *Report of the commissioner of Crown Lands, 1865* (1866), App. 25a; ibid., 1866 (1867), App. 24; and "Report on the Country Between Lake Superior and Lake Winnipeg, by Robert Bell," in GSC *Report of Progress, 1872-73* (1873), pp. 87-111. Bell visited the Lake Superior mining region, reporting much activity in exploration, prospecting, and mining. See also Crown Lands Department, vol. 4, "Register of Mining Licenses, Lake Superior, 1869-73"; and *Weekly Herald and Lake Superior Mining Journal,* 17 June, 23 September 1882.
59. See Thomas Macfarlane, "Silver Islet," AIME *Trans* 8 (1879-80):226-53. See also "On the Laurentian, Huronian, and Upper Copper-bearing rocks, and the Economic Minerals of Michipicoten Island and the East Shore of Lake Superior, by T. Macfarlane," in GSC *Report of Progress, 1963-66* (1867) [hereafter "Macfarlane's Report, 1863-66"]; and Arthur, ed., *Thunder Bay District,* p. 105.
60. "On the Geology and Economic Minerals of the North Shore of Lake Superior with Remarks on Copper Ores and their Treatment, by W.E. Logan," in GSC *Report of Progress, 1846-47* (1847); and "Report on Mines and Mining on Lake Superior . . ., by E.D. Ingall," in GSC *Summary Report, 1887 and 1888* (1888) [*hereafter, "Ingall's Report, 1888"*].
61. See Arthur, ed., *Thunder Bay District,* introduction and pp. 83-173.
62. Elwood S. Moore, *American Influence in Canadian Mining* (Toronto: University of Toronto Press, 1941), foreword.
63. *Weekly Herald and Lake Superior Mining Journal,* 23 September 1882.

64. Ibid., 17 June 1882; 12 August 1882.
65. *Thunder Bay Sentinel,* 29 July 1875; and Notes 58 and 60. See Alexander Stewart, *Thunder Bay Silver Mining Company Report* (Montreal: J. Starke, 1874).
66. "Ingall's Report, 1888," Part 1B, pp. 51-53; and *Ontario Commission on Mineral Resources,* pp. 34, 197.
67. *Thunder Bay Sentinel,* 5 August 1882.
68. Macfarlane, "Silver Islet," pp. 226-53; "Bell's Report for 1872-73," p. 108; and "Ingall's Report, 1888," Pt. 1B, pp. 14-40.
69. Ibid., Pt. 1B, pp. 51-53; and Macfarlane, "Silver Islet," pp. 42-44. Two pamphlets on Silver Islet also provide useful information. See A.H. Sibley, *Report on Mining on the North Shore of Lake Superior, by A.H. Sibley* (1873); and Janey C. Livingstone, *Historic Silver Islet. The Story of a Drowned Mine* (Fort William: Times-Journal Press, n.d.). J. Livingstone was the daughter of the government man at Silver Islet. See also Helen Moore Strickland, *Silver Under the Sea* (Cobalt, Ontario, 1979). The mine operated under the direction of a subsidiary of Ontario Mineral Lands called Silver Islet Mining Company.
70. See various issues of the *Thunder Bay Sentinel,* for example, 5 and 8 October 1880. For an idea of costs, see Cross Family Papers, Correspondence, "Supplies Used in the Mine the Month of May, 1878," PAO.
71. *Thunder Bay Sentinel,* 28 October 1875, "A Visit to Silver Islet." See also Macfarlane, "Silver Islet;" Archibald Blue, "The Story of Silver Islet," in Ontario Bureau of Mines, *Report, 1892* (1892), pp. 125-57; Cross Family Papers, J.S. Cross, "Silver Islet Social Conditions" (typescript, 1896), p. 1; and Livingstone, *Historic Silver Islet.* According to J.S. Cross, the hospital was located above the reading room, and the saloon occupied one room in one of the boarding houses. The men were limited to three drinks each per day on the Islet, depending upon their conduct. It seems the company preferred to control access to alcohol because of the "whiskey peddlars" who roamed the region. The jail house was a substantial building, with five cells and a bedroom for the jailer. Of some interest for social information on these mining communities are the Silver Islet company store ledgers (scraps) in the Winnifred Philpot Papers, PAC.
72. Toronto *Globe,* 11 and 21 October 1865; 19 February 1874; and Cross Family Papers, introduction to the collection.
73. Blue, "The Story of Silver Islet," p. 149; and a pamphlet by Peter McKellar, *Mining on the North Shore of Lake Superior* (n.p. 1874), pp. 10-15.
74. See Robert M. Vogel, "Tunnel Engineering: A Museum Treatment," Paper 41, USM *Bulletin* 240 (1964):210; E. Gybbon Spillsbury, "Rock-Drilling Machinery," AIME *Trans* 3 (1874-75):144-50; and Larry D. Lankton, "The Machine *Under* the Garden: Rock Drills Arrive at the Lake Superior Copper Mines, 1868-1883," *Technology and Culture* 24, 1 (1983):1-37.
75. Arthur, ed., *Thunder Bay District,* p. 147.
76. Cross Family Papers, Frue *Letterbook,* 1871-1875, Frue to Trowbridge, 19 February 1872; Frue to Sibley, 19 February 1872; and Frue to Wm. Savard, 30 July 1873; Gates, *Michigan Copper and Boston Dollars,* p. 113. According to Gates, this was Michigan's first major miners' strike.
77. Cross Family Papers, Frue *Letterbook,* 1871-1875, Frue to Wm. Savard, 30 July 1873. See also Sibley, *Report on Mining on the North Shore of Lake Superior.*
78. It would be unwise to place too great an emphasis on the underground development work, however, especially for the first few years, when the company's own records indicate that ninety cents of every dollar was spent on the "permanent improvements" at the surface. Cross Family Papers, Frue *Letterbook,* 1871-1875, Frue to Duffield, 2 November 1872.
79. Ibid., Frue to Trowbridge, 30 July 1873.
80. *Thunder Bay Sentinel,* 2 September 1875, 21 October 1875, and 8 October 1880.
81. "Records of Mines and Mineral Statistics Compiled by Charles Robb," Table I, Ontario, in GSC *Report of Progress, 1871-72* (1872), p. 147; McKellar, *Mining on Lake Superior,* pp. 11-12; and *Thunder Bay Sentinel,* 28 October 1875.
82. Ibid., 20 July 1875, 14 September 1876.
83. Thomas Macfarlane was superintendent of the works. See W.E. Courtis, "The Wyandotte Silver Smelting and Refining Works," AIME *Trans* 2 (1873-74):89-100. (Courtis was manager of the Duncan Mine.) See also, Cross Family Papers, Smelting Statements, 1872-79.
84. See "Ingall's Report, 1888," p. 38.
85. Detailed in Fraser and Chalmers, Co., *Catalogue No. 3* (Chicago, 1888?), pp. 3-5.

86. Ibid., "List of Companies Supplied – Crushing Mills," p. 111. For details on experimentation with the vanner, see *Sentinel,* 28 October 1875; 11 May 1876 (reprint from the *Western Daily News,* Plymouth, England); 21 September 1876 (reprint from the *Denver Mining Review*); and Cross Family Papers, Frue *Letterbook,* 1871-75, Frue to Walter McDermott, 19 November 1873. See also Canada, Patent No. 3,974 (26 October 1874), extended, No. 10,487, 27 September 1879.
87. Blue, "The Story of Silver Islet," pp. 125-57.
88. "Ingall's Report, 1888," pt. 1B, pp. 63, 70, 74.
89. *Thunder Bay Sentinel,* 11 May 1876.
90. Ibid., 28 October 1875 (reprint of a paper read at New Haven, Conn., by Walter McDermott). John Adams, first graduate of the Columbia School of Mines, served as agent for the Frue Vanner. For details consult Clark C. Spence, *Mining Engineers and the American West: The Lace-Boot Brigade, 1849-1933* (Ithaca: Cornell University Press, 1958), p. 247. See Fraser and Chalmers, Co., *Catalogue No. 3,* "List of Companies Supplied with Frue Vanning Machines," pp. 97-100; and "Frue Vanning Machine and Concentrator," *Mining World and Engineering Record* (11 March 1876).
91. *Engineering* 31 (10 December 1884):542 (advertisement), 40 (5 March 1886):222, 228 (advertisement).
92. *Ontario Commission on Mineral Resources,* pp. 510, 511; and "Ore Testing Works and Recent Additions to the Assay Department of the School of Mines," *The School of Mines Quarterly* 6, 3 (1885):224-35.
93. H.S. Munroe, "The Losses in Copper Dressing at Lake Superior," AIME *Trans* 8 (1879-80):409-51. The Frue Vanner was tested at the Osceola mill (p. 442).
94. "Ingall's Report, 1888," Pt. 1B, pp. 32-33.
95. Ibid., pp. 56-63; and *Thunder Bay Sentinel,* 13 September 1878, 6 August 1880.
96. *Ontario Commission on Mineral Resources,* pp. 198-99.
97. Ibid., pp. 191-204; and Livingstone, *Historic Silver Islet,* p. 21.
98. Cross Family Papers, *Silver Islet Time Book,* 1882-1884, and 1884-1886; and Silver Islet Consolidated Mining and Lands Co., *Report of the Directors to Stockholders* (1879, 1884).
99. This theme is well developed in M.A. Wright, "The Canadian Frontier"; Dennis Watson, "Frontier Movement and Economic Development in Northeastern Ontario, 1850-1914" (Master's thesis, University of British Columbia, 1971); and W.T. Jestin, "Provincial Policy."
100. "Ingall's Report, 1888," Pt. 1A., p. 12; and *Ontario Commission on Mineral Resources,* pp. 98, 102-3.
101. *Ontario Commission on Mineral Resources,* pp. 60-61, 103; "Ingall's Report, 1888," Pt. 1A, pp. 11-12; and "Macfarlane's Report, 1863-66."
102. *Ontario Commission on Mineral Resources,* p. 147, 29, 30; "Ingall's Report, 1888," Pt. 1A, p. 13; and Department of Indian Affairs, *Annual Report, 1885* (1885), "Reports of Superintendent and Agents." The agent here reported that Indians had found work at the local mines and also winter employment in the United States. I am grateful to A.J. Ray for bringing this information to my attention.
103. *Ontario Commission on Mineral Resources,* p. 142-43, 123-27, 144-45.
104. For good descriptions of the mining operations in the Rabbit Mountain region, see "Description of the Various Mines in Operation in the Rabbit Mountain Region . . . , by A.L. Russell," in Ontario, *Report of the Commissioner of Crown Lands, 1885* (1886), App. 20; and T.A. Keefer Papers, "Speech, 1885."
105. *Ontario Commission on Mineral Resources,* pp. 192-94, 196, 198-204; "Ingall's Report, 1888," pp. 72-94; and Department of Indian Affairs, *Annual Report, 1885* (1885), "Reports of Superintendent and Agents." The superintendent reported that Indians in this region were more interested in working in the mines (and on the railway) than in farming.
106. See testimony of its mining captain, Thomas Hooper, *Ontario Commission on Mineral Resources,* pp. 198-99. Details on the Beaver Mine appeared in the *Algoma Miner and Weekly Herald,* 27 August 1887. See also "Report of the Gold Mines of the Lake of the Woods, by E. Coste," in GSC *Report of Progress, 1882-84* (1885) [hereafter, "Coste's Report, 1882-84"]; and "The Port Arthur Mines," *Canadian Mining Review* 8, 1 (1889):11-12.
107. *Canadian Mining Review* 9, 7 (1890):102; Ontario Bureau of Mines, *Annual Report, 1891* (1892), p. 227; and ibid., 1893 (1894), pp. 19, 52.
108. *Ontario Commission on Mineral Resources,* pp. 201-2; and Department of Mines, RG 87, Mineral Statistics, vol. 31, file 129, PAC.
109. See Morris Zaslow, "The Ontario Bound-

ary Question," in *Profiles of a Province,* ed. Edith G. Firth (Toronto: Ontario Historical Society, 1967), pp. 107-17.

110. McKellar's presentation to the Canadian Institute (Winter, 1874) is summarized in the *Algoma Miner and Weekly Herald,* 24 January 1883. See also ibid., 28 May 1883.

111. See Note 58; and Department of Crown Lands, RG 1, G-1, vol. 1, "Henry Lloyd's Mining Lands, 1871-74." (The correspondence concerns mineral lands in the Rat Portage, Rainy River region.) For background information, consult Morris Zaslow, *The Opening of the Canadian North, 1870-1914* (Toronto: McClelland and Stewart, 1971), chapter 7; and Zaslow, "The Ontario Boundary Question," pp. 107-17.

112. *Ontario Commission on Mineral Resources,* p. 241. See also *Algoma Weekly Herald,* 3 July 1884.

113. Ibid., 108-9, 116-19; and "Coste's Report, 1882-84," pp. 11-19.

114. Fraser and Chalmers Co., *Catalogue No. 3,* p. 102.

115. A thorough and highly readable historical account is contained in *Ontario Commission on Nickel,* pp. 30-56. See also *Ontario Commission on Mineral Resources,* pp. 103-5, 30-56, 88-89; and Watson, "Frontier Movement and Economic Development."

116. *Ontario Commission on Mineral Resources,* testimony of John Babcock, p. 105, James Stobie, pp. 53, 143. See also the autobiography of a successful Sudbury pioneer and mine broker, Æneas McCharles, *Bemocked of Destiny* (Toronto: William Briggs, 1908). McCharles bequeathed prize money to be awarded by the University of Toronto to Canadian inventors of mining equipment. *Bemocked* is curiously rambling and contains almost no references to the author's actual mining experiences. Sudbury's early beginnings are detailed in Gilbert A. Stelter, "The Origins of a Company Town: Sudbury in the Nineteenth Century," *Laurentian University Review* 3, 3 (1971):16-17. See also Thomas W. Gibson, *Mining in Ontario* (Toronto: King's Printer, 1937), pp. 96ff.

117. Ontario Bureau of Mines, *Annual Report, 1893* (1894), pp. 55-61; Nelles, *The Politics of Development,* pp. 25, 90; and Gilbert A. Stelter, "Community Development in Toronto's Commercial Empire: The Industrial Towns of the Nickel Belt, 1883-1931," *Laurentian University Review* 6, 3 (1974):3-53.

118. *Ontario Commission on Mineral Resources,* pp. 433-35; and Edward D. Peters, Jr., "The Sudbury Ore Deposits," AIME *Trans* 18 (1889-90):278-89.

119. Peters, *Modern American Methods of Copper Smelting;* Peters's testimony, *Ontario Commission on Mineral Resources,* pp. 103, 232, 310, 404; and F.F. Andrews's testimony, ibid., pp. 104, 232, 241, 311, 426.

120. *Ontario Commission on Mineral Resources,* p. 104.

121. Its use at Sudbury is described in Peters, "The Sudbury Ore Deposits," pp. 286-87; *Ontario Commission on Mineral Resources,* pp. xxii-xxiii, 377, 378-80; and Fraser and Chalmers Co., *Catalogue No. 3,* pp. 109, 27-28.

122. See Peters, "The Sudbury Ore Deposits," 284-86.

123. The most recent study of this famous company is Wallace Clement, *Hardrock Mining: Industrial Relations and Technological Changes at INCO* (Toronto: McClelland and Stewart, 1981).

124. See Arthur, *Thunder Bay District,* introduction; Sanford Fleming, *Report on Surveys and Preliminary Operations on the Canadian Pacific Railway up to January, 1877* (Ottawa: Queen's Printer, 1877); and Peters, "The Sudbury Ore Deposits," pp. 287-88. Pennsylvania coke was imported via the Great Lakes and the Algoma branch of the Canadian Pacific Railway at an "excellent" rate. It required one ton of coke to process seven or eight tons of copper ore – this was considered a favourable ratio.

NOTES TO CHAPTER FIVE

1. An important work on this topic is J.H. Richards, "Population and the Economic Base in Northern Hastings County, Ontario," *Canadian Geographer* 11 (1958):22-31. See also M.L. Bladen, "Construction of Railways in Canada to the Year 1885," pt. 1, *Contributions to Canadian Economics* 1, 5 (1932):43-60.

2. The most comprehensive source for these developments is George W. Spragge, "Colonization Roads in Canada West, 1850-1867," *Ontario History* 49 (1957):1-17. See also Florence B. Murray, "Agricultural Settlement on the Canadian Shield," in *Profiles of a Province,* ed. Edith G. Firth (Toronto: Ontario Historical Society, 1967), pp. 178-86; and J.H. Richards, "Lands and Pol-

icies: Attitudes and Controls in the Alienation of Lands in Ontario during the First Century of Settlement," *Ontario History* 50 (1958):192-209.

3. See Elwood S. Moore, *American Influence in Canadian Mining* (Toronto: University of Toronto Press, 1941), pp. 24-26.
4. Ontario, Royal Commission on Mineral Resources, *Report* (Toronto: Warwick and Sons, 1890) [hereafter, *Ontario Commission on Mineral Resources*], pp. 321-23; and W.G. Miller, "Mines and Mining," in *Canada and Its Provinces,* eds. A. Shortt and A. Doughty, vol. 18, *Ontario,* pt. 2 (Toronto: Glasgow, Brook, and Co., 1914), pp. 613-45. Also of interest is Robin Hood, "The History of [Iron] Mining [at Marmora]" (unpublished manuscript prepared for the Crowe Valley Conservation Authority, c. 1975).
5. Rita Michael, "Ironworking in Upper Canada: Charles Hayes and the Marmora Works," *CIM Bulletin* 76, 849 (1983):132-35.
6. Ibid., p. 134.
7. Thomas Hetherington, Anthony Manahan, and Peter McGill. Upper Canada, *Statutes,* 1 Wm. 4, 1830, c. 11, "An Act to Incorporate . . . The Marmora Foundry Co." See Hood, "The History of Mining," pp. 18-22.
8. Richards, "Northern Hastings County," p. 29. See, for example Bruce E. Seely, "Blast Furnace Technology in the Mid-19th Century: A Case Study of the Adirondack Iron and Steel Company," *IA* 7, 1 (1981):27-54.
9. For a detailed account, see *Ontario Commission on Mineral Resources,* pp. 323-25, 389.
10. See Miller, "Mines and Mining," pp. 113-15.
11. *Ontario Commission on Mineral Resources,* pp. 323-25.
12. See account in Hood, "History of Mining," pp. 28-30.
13. *Ontario Commission on Mineral Resources,* pp. 321-22.
14. For a detailed account, see ibid., pp. 321-23.
15. Ibid., p. 322; Province of Canada, *Statutes,* 16 Vict., 1853, c. 253, "An Act to Amend the Act of Upper Canada, incorporating the Marmora Foundry Co." Peter McGill was also a director of this company. See also Ontario Bureau of Mines, *Report, 1892,* p. 21.
16. *Ontario Commission on Mineral Resources,* p. 322; and see also R.R. Potter, "The Woodstock Ironworks, Carleton County, New Brunswick" (paper read before the Annual Conference of Metallurgists, Ottawa, August 1982).
17. Hood, "History of Mining," p. 31; and Tom Naylor, *The History of Canadian Business, 1867-1914,* vol. 1 (Toronto: James Lorimer, 1975), p. 53.
18. Province of Canada, *Statutes,* 29-30 Vict., 1866, c. 103; and 33 Vict., 1869, c. 38. The Cobourg and Peterborough Railway Co. in the previous year had been authorized to construct a tramway or railway from the Marmora Iron Works to the Trent River or to Rice Lake (ibid., 29 Vict., 1865, c. 81).
19. "Notes on the Iron Ores of Canada and Their Development, by B.J. Harrington," in Canada, Geological Survey of Canada [GSC], *Report of Progress, 1873-74* (1874) [hereafter, "Harrington's Report, 1873-74"], p. 257, "The amount of iron produced here in the season 1872-73 was 30,000 tons, approximately one-half the total for the Dominion."
20. Moore, *American Influence,* p. 24; and "Harrington's Report, 1873-74," pp. 244, 257.
21. *Ontario Commission on Mineral Resources,* p. 323. The invention likely was "Smith's Process for Heating Metals with Oil as Fuel and Smelting Ores," patented in Canada by Ananian Smith, a Clifton, Ontario, machinist (10 January 1876, No. 5,554). No details concerning this experiment or the reasons for its failure have come to light.
22. Toronto *Globe and Mail,* 8 November 1938, cited in Hood, "History of Mining," p. 38. The orebody, however, was reworked many times in the twentieth century.
23. Descriptions of the Chaffey, Yankee, and Dalhousie mines are contained in "Progress Report of Exploration and Surveys in the Counties of Leeds, Frontenac and Lanark, with Notes on the Gold of Marmora . . . by H.G. Vennor," in GSC, *Report of Progress for 1871-72* (1872) [hereafter, "Vennor's Report, 1871-72"]. Additional details on the entrepreneurial activities of the Chaffey family are reported in Edward F. Bush, "Commercial Navigation on the Rideau Canal, 1832-1961," *History and Archaeology* 54 (Ottawa: Parks Canada, 1981), especially p. 118.
24. Ibid.
25. "Report of Explorations and Surveys in the Counties of Addington, Frontenac, Leeds and Lanark, by H.G. Vennor," in GSC, *Report of Progress for 1872-73* (1873) [hereafter, "Vennor's Report, 1872-73"], pp. 176, 178.
26. *Ontario Commission on Mineral Resources,* pp.

138-39.

27. "Vennor's Report, 1872-73," p. 177.
28. "Harrington's Report, 1873-74," p. 245.
29. GSC, *Mineral Statistics and Mining, 1887* (1889), p. 34; and *Mineral Statistics and Mining, 1888* (1889), p. 42.
30. Bush, "Commercial Navigation on the Rideau Canal", p. 124.
31. *Ontario Commission on Mineral Resources,* pp. 132-35.
32. Ibid., pp. 133-34.
33. Ibid., pp. 132-35.
34. Ibid.; and "Report of Observations in 1883 on Some Mines and Minerals in Ontario . . . by C.W. Willimott," in GSC, *Report of Progress, 1882-84* (1885) [hereafter, "Willimott's Report, 1882-84"], p. 11ff.
35. RG 87, Department of Mines, Mineral Statistics [hereafter, Department of Mines, Mineral Statistics], vol. 18, file 82, PAC.
36. Ontario, *Report of the Commissioner of Crown Lands, 1884* (1885), App. No. 20, pp. 123-29.
37. *Ontario Commission on Mineral Resources,* p. 134.
38. Ontario, *Report of the Commissioner of Crown Lands, 1884* (1885), App. No. 20, pp. 123-29; and *Ontario Commission on Mineral Resources,* p. 129.
39. *Ontario Commission on Mineral Resources,* p. 135.
40. *Engineering and Mining Journal* 36 (1885):345; and Department of Mines, Mineral Statistics, vol. 18, file 82, PAC.
41. Department of Mines, Mineral Statistics, vol. 18, file 82, PAC.
42. *Ontario Commission on Mineral Resources,* p. 136.
43. Ibid., pp. 128-29.
44. Department of Mines, Mineral Statistics, vol. 18, file 82, PAC; and *Ontario Commission on Mineral Resources,* pp. 141, 146. Clymo also worked at the Ramsay (lead) Mine.
45. See historical sketch contained in Ontario, *Report of the Commissioner of Crown Lands, 1884* (1885), App. No. 21, pp. 31-32; and *Ontario Commission on Mineral Resources,* pp. 127-29.
46. Moore, *American Influence,* pp. 24-25; H.V. Nelles, *The Politics of Development, Forest, Mines and Hydro-Electric Power in Ontario, 1849-1941* (Toronto: Macmillan, 1975), pp. 24, 62; and Morris Zaslow, *The Opening of the Canadian North, 1870-1914* (Toronto: McClelland and Stewart, 1971), chapter 7.
47. For details on the testing of new properties, see E.J. Chapman, "Report of Professor E.J. Chapman on Mineral Locations in North Hastings, Ontario" (Toronto, 1881).
48. Richards, "Northern Hastings County," p. 29; and Thomas W. Gibson, *Mining in Ontario* (Toronto: King's Printer, 1937), pp. 118-19.
49. Gibson, *Mining in Ontario,* pp. 79-84, 120; Moore, *American Influence,* pp. 25-26; Nelles, *The Politics of Development,* pp. 24-25, 90; and Gilbert A. Stelter, "The Origins of a Company Town: Sudbury in the Nineteenth Century," *Laurentian University Review* 3, 3 (1971):1-10.
50. E.J. Chapman, "On the Wallbridge Hematite Mine, as Illustrating the Stock-Formed Mode of Occurrence of Certain Ore Deposits," *Proceedings and Transactions of the Royal Society of Canada* 4 (1885):23-26.
51. See testimony of John Stewart, an Ottawa mining engineer who had worked at the Wallbridge Mine, before the Royal Commission (*Ontario Commission on Mineral Resources,* pp. 132-33).
52. Chapman, "On the Wallbridge Hematite Mine," pp. 23-26.
53. *Ontario Commission on Mineral Resources,* p. 131.
54. See Gibson, *Mining in Ontario,* p. 114.
55. Appendix C, on phosphate of lime, in *Ontario Commission on Mineral Resources,* pp. 167-81.
56. Except where indicated otherwise, the following description of apatite-mining has been taken from "Progress Report of Exploration and Surveys in the Rear Portion of Frontenac and Lanark Counties, together with Notes on Some of the Economic Minerals of Ontario, by H.G. Vennor," in GSC, *Report of Progress, 1874-75* (1876) [hereafter, "Vennor's Report, 1874-75"], pp. 110, 170-71.
57. Bush, "Commercial Navigation on the Rideau Canal," p. 127, cites the Tett Family Papers, Queen's University Archives, Kingston, Ontario.
58. *Ontario Commission on Mineral Resources,* p. 174.
59. Ibid., p. 170.
60. For a thorough overview of apatite operations in Canada in 1880s, see GSC, *Mineral Statistics and Mining, 1890* (1891), pp. 153-61.
61. For details, see James Foxton's testimony before the Royal Commission (*Ontario Commission on Mineral Resources,* pp. 168, 171-72).
62. Ibid.
63. Ibid., p. 176.

64. D.D. Hogarth, "Apatite in the Ottawa and Kingston Areas: A Flashback" (paper read before the symposium, Historic South Shield Mining Resources and Future Public Management," Kempville, Ontario, 23 June 1976), p. 8.
65. "Vennor's Report, 1872-73," p. 179.
66. "Record of Mines and Mineral Statistics, by Chas. Robb," in GSC, *Report of Progress, 1871-72* (1872), pp. 146-54.
67. "Notes on the Phosphate of Lime and Mica found in North and South Burgess and North Elmsley, Ontario, by G. Broome," in GSC, *Report of Progress, 1870-71* (1872), p. 321.
68. "Vennor's Report, 1872-73," pp. 178-79.
69. Moore, *American Influence*, p. 260.
70. Well-documented in *Ontario Commission on Mineral Resources*, p. 149.
71. Ibid., p. 148; and "Register of Mining Licenses 1877-1910," Canada Company Papers, PAO. The Sydenham Mining and Mica Company in 1888 paid a royalty of $1.25 per 2,000 lbs. extracted.
72. For a fuller account of the markets and changing market situations, see GSC, *Mineral Statistics and Mining, 1890* (1891), pp. 102-5.
73. "Head's Machine for Grinding Mica," patented 20 December 1886, Canada Patent No. 25,593.
74. Ontario Bureau of Mines, *Report, 1893*, pp. 192-96.
75. "On the Geology and Economic Minerals of the North Riding of the County of Hastings, Townships of Elzevir, Madoc, Marmora, Lake, and Tudor, by T. Macfarlane," in GSC, *Report of Progress, 1863-66* (1867), pp. 103-5; and "On the Geology and Economic Minerals of Parts of Hastings, Addington, and Peterborough Counties, by H.G. Vennor," in GSC, *Report of Progress 1866-69* (1870) [hereafter "Vennor's Report, 1866-69"], p. 162.
76. Charles Robb, "Report on the Frontenac Lead Mine" (n.p., 1868); J.W. Dawson, "Report on the Frontenac Lead Mine" (n.p., 1868); and E.J. Chapman, "Report on the Frontenac Lead Mine" (n.p., 1868), all PAO.
77. Dawson, "Report on the Frontenac Lead Mine."
78. Robb, "Report on the Frontenac Lead Mine."
79. "Vennor's Report, 1866-69," pp. 164-65; and "Record of Mines and Mineral Statistics, 1871-72," p. 147.
80. Chapman, "Report on the Frontenac Lead Mine."
81. *Ontario Commission on Mineral Resources*, p. 145. See Forbes Galena Mines Papers, 1868-1951, Gillies Brothers Lumber Company Records, PAO, for a description of the attempts in 1911 and the 1920s to resurrect mining operations at this site.
82. Province of Canada, *Statutes*, 22 Vict., 1859, c. 112, "An Act to Incorporate the Ramsay Lead Mining and Smelting Company," amended 25 Vict., 1862, c. 75.
83. See testimony of Charles Clymo, the Cornishman who worked this mine (*Ontario Commission on Mineral Resources*, p. 146).
84. Ibid., p. 145.
85. "Vennor's Report, 1874-75," p. 163.
86. "Willimott's Report, 1882-84," p. 9.
87. "Vennor's Report, 1866-69," p. 165.
88. Gibson, *Mining in Ontario*, pp. 3-4. The mineral occurred in the form of free gold in quartz and mispickel, or "fool's gold."
89. See Marilyn G. Miller, "Small Scale Mining in the South Shield Region of Eastern Ontario" (unpublished study prepared for the Ontario Ministry of Culture and Recreation and Ministry of Natural Resources, 12 March 1976), pp. 57-60; Canada, *Report of the Commissioner of Crown Lands*, 1867 (1868), p. ix; Ontario Bureau of Mines, *Report, 1893*, pp. 53-61. See also correspondence 21 February 1867, 18 and 29 March 1967, A.N. Buell Papers, PAO.
90. "Improvements in the Construction of Amalgamating Tables for Separating Gold from Pulverized Ore," 7 May 1867, Canada Patent No. 2,325 (Robert Smith, Toronto Merchant); "Harrison and Caldwell's Improved Amalgamator," 12 December 1869, No. 2,901 (Robert Harrison and William Macauley were Toronto miners); "Electrical Cylinder for Separating the Metals and Auriferous and Argentiferous Rocks," 28 August 1868, No. 2,745 (John Neale was a Frontenac County miner); "O'Brien's Gold and Silver Amalgamator and Grinding Pan," 14 January 1869, No. 2,963 (G. O'Brien was an Eldorado innkeeper); "Dunstan's Revolving Cylinder-Furnace," 11 January 1872, No. 1,279 (J.H. Dunstan was manager of the Dean and Williams Mine); "Process for Extracting Gold from Ores," 31 August 1872, No. 1,602 (John Cull, esq., Toronto); "Howell's Process and Apparatus for Treating Refractory Ores," 12 January 1873, No. 7,994, reissued 1878, No. 8,293 (Henry F. Howell, esq., of St.

Catharines and Sarnia, who also patented a number of inventions for the petroleum industry); and "Long's Patent Gold Separator and Filter," 29 October 1868, No. 2,803 (Asa Long was a local millwright).

91. "Vennor's Report, 1866-69," p. 169.
92. Gibson, *Mining in Ontario,* p. 6.
93. Malcolm Wallbridge, ed., *The Diary of Thomas Nightengale, Farmer and Miner, 1867-1871* (Picton: Picton Gazette, 1967).
94. Ibid., p. 43 (entries for 27, 28, and 30 November 1868). 29 October 1868, Canada Patent No. 2,803.
95. See "Vennor's Report, 1866-69," p. 165ff; and "Vennor's Report, 1871-72," p. 132.
96. "Vennor's Report, 1871-72," p. 132ff.
97. Ibid.
98. MG III, 40, Macdonell and Foster Papers, Letterbook, 1870-1871, PAC [hereafter, Macdonell and Foster Papers]. Other aspects of the firm's dealings are discussed in Chapter 3.
99. Ibid., Fred Foster to Chief Administrator, Nova Scotia Mines, 4 February 1871.
100. Patented 10 October 1870, Canada Patent No. 735.
101. Macdonell and Foster Papers, Fred Foster to Dean and Williams, 29 December 1870.
102. Ibid., Fred Foster to I.W. Forbes, 11 January 1871.
103. Ibid.
104. Ibid.
105. Ibid., Fred Foster to I.W. Forbes, 19 and 26 March 1871.
106. Ibid., Fred Foster to I.W. Forbes, 2 February 1871; and Fred Foster to Macdonell, 21 March 1871. "Forbes' Art of Cutting, Grinding, Cleaning, and Separating and Securing Crushed or Pulverized Ore or Metals," patented 13 January 1871, Canada Patent No. 781.
107. "Vennor's Report, 1871-72," pp. 134-36; RGIG. Vol. 7, Department of Crown Lands, "Mining Inspector's Report, Madoc Mining Division," 16 December 1873, PAO; and Ontario, *Report of the Commissioner of Crown Lands, 1873* (1874).
108. Ontario, *Report of the Commissioner of Crown Lands, 1873* (1874), pp. 32-37. Patented 11 January 1872, Canada Patent No. 1,279.
109. Wallbridge, *Diary of Thomas Nightengale,* 17 February and 1 March 1871.
110. Gibson, *Mining in Ontario,* p. 7.
111. 29 January 1873, Canada Patent No. 2,015, and 3 February 1873, Canada Patent No. 2,026.
112. In addition to "Vennor's Report, 1871-72," see Wallbridge, *Diary of Thomas Nightengale,* for a running commentary on goldmining in the area.
113. "Vennor's Report, 1871-72," p. 40.
114. Ibid., pp. 138-40.
115. *Ontario Commission on Mineral Resources,* pp. 106-7.
116. "Vennor's Report, 1871-72," pp. 138-40.
117. Ontario, *Statutes,* 36 Vict., 1873, c. 109; amended 39 Vict., 1875-76, c. 89.
118. See "Prospectus of the Canada Consolidated Gold Mining Company," Misc. Pamphlets, PAO.
119. Brief histories of this operation are contained in "Willimott's Report, 1882-84," pp. 13-14; and Moore, *American Influence,* p. 26; and *Ontario Commission on Mineral Resources,* pp. 106-8. A complete description of the surface workings is contained in R.P. Rothwell, "The Gold-Bearing Mispickel Veins of Marmora, Ontario, Canada." AIME *Trans* 9 (1880-81):409-20.
120. H.S. Munroe, "The New Dressing Works of the St. Joseph Lead Company, at Bonne Terre, Missouri," *Canadian Mining Review* 9, 9 (September 1890):127. For more details refer to Chapter 2.
121. *Ontario Commission on Mineral Resources,* pp. 107-8. For an excellent overview of Western amalgamation techniques see A.D. Hodges, Jr., "Amalgamation at the Comstock Lode, Nevada: A Historical Sketch of Milling Operations at Washoe, and an Account of the Treatment of Tailings at the Lyon Mill, Dayton," AIME *Trans* 19 (1890-91):195-231.
122. *Ontario Commission on Mineral Resources,* p. 106-8. For details on the same process applied to a different site, see George W. Small, "Notes on the Stamp Mill and Chlorination Works of the Plymouth Consolidated Gold Mining Company, Amadore County, California," AIME *Trans* 15 (1886-87):305-8.
123. Ontario Bureau of Mines, *Report, 1893,* pp. 55-61.

NOTES TO CHAPTER SIX

1. "On the Geology of the Peninsula Bounded by Lakes Huron, St. Clair and Erie, by A. Murray," in Geological Survey of Canada [GSC], *Report of Progress, 1850-51* (1852). Murray noted that the bituminous oil was

frequently collected on cloths and used as a remedy for wounds; and "On . . . Mineral Springs . . . Minerals and Metallic ores . . . , by T.S. Hunt," in GSC, *Report of Progress, 1848-49* (1850).

2. For a general discussion of the early attempts, see Edward Phelps, "John Henry Fairbank of Petrolia (1831-1914)" (M.A. thesis, University of Western Ontario, 1965), p. 18. Phelps cites an earlier work by Robert B. Harkness, "Makers of Oil History, 1850-1880" (unpublished manuscript, Regional History Collection, University of Western Ontario Library, n.d.), pp. 20-21.
3. Province of Canada, *Statutes,* 18 Vict., 1854, c. 77. See discussion on the Tripp enterprise in Phelps, "John Henry Fairbank," p. 18.
4. See Phelps, "John Henry Fairbank," p. 18 and fn. 7; and J.J. Brown, *Ideas in Exile: A History of Canadian Invention* (Toronto: McClelland and Stewart Ltd., 1967), p. 64.
5. See biographical essay on Gesner by Loris Russell, in *Dictionary of Canadian Biography,* vol. 9 (Toronto: University of Toronto Press, 1977), pp. 308-12, with a select list (p. 312) of Gesner's writings. Of special importance is Arnold R. Dawn's examination of the development of the illuminating industry that conditioned the emergence of the petroleum industry and the impact of the coal oil industry on petroleum refining and marketing. See Arnold R. Dawn, "The Illumination Revolution and the Rise of the Petroleum Industry, 1850-1863" (Ph.D. dissertation, Columbia University, 1957); Ian Sclanders, "He Gave the World a Brighter Light," *IOR* 39, 1 (1955):23-25; and Harold F. Williamson and Arnold R. Daum, *The American Petroleum Industry,* vol. I, *The Age of Illumination, 1859-1899* (Evanston: Northwestern University Press, 1959), pp. 43-60.
6. Phelps, "John Henry Fairbank," pp. 18-19.
7. Ibid., p. 19; and Fergus Cronin, "North America's Father of Oil," *IOR* 39, 2 (1955): 17-20.
8. Ontario, Royal Commission on Mineral Resources, *Report* [hereafter, *Ontario Commission on Mineral Resources*] (Toronto: Warwick and Sons, 1890), pp. 158-59. See also Norman R. Ball, "Petroleum Technology in Ontario during the 1860s" (M.A. thesis, University of Toronto, 1972), chapter 7, which he devotes to a discussion of surface oil and the variety of surface wells dug in the Enniskillen field. Ball cites a number of contemporary newspaper accounts collected and preserved in scrapbooks by George and Leslie Smith of Sarnia.
9. Ball, "Petroleum Technology," pp. 29-30, cites *Canadian News,* 27 September 1866, p. 194.
10. Phelps, "John Henry Fairbank," p. 19.
11. *Ontario Commission on Mineral Resources,* p. 159.
12. P.W. Andrews, "The Canadian Mineral Industry and the Economy," Canada Department of Energy, Mines and Resources, *Mineral Bulletin* MR128 (1972):64.
13. Cronin, "North America's Father of Oil," p. 20.
14. *Ontario Commission on Mineral Resources,* p. 160.
15. The role of Hunt in developing the anticlinal theory in connection with the accumulation of gas and petroleum is described in Morris Zaslow, *Reading the Rocks: The Story of the Geological Survey of Canada, 1842-1972* (Toronto: Macmillan, 1975), p. 76; and Robert Bell, "The Petroleum Fields of Ontario," *Proceedings and Transactions of the Royal Society of Canada* 5 (1888);102-4. Hunt's paper on the subject was published in the Montreal *Gazette,* 1 March 1861. See also "Report on Natural Gas and Petroleum in Ontario Prior to 1891, by H.P.H. Brumell," in GSC, *Summary Report,* 1891 (1892) [hereafter, "Brumell's Report on Natural Gas and Petroleum"], pp. 10-21, for an historiography of theories on the origins of hydrocarbons. Hunt and Bell made it clear that drillers would have to be guided by the data supplied by the logs of wells drilled in various parts of the possible oil region. Hunt performed a valuable service to the industry by gathering all extant boring logs and publishing these in, "On the Geology and Mineralogy of Laurentian Limestones; Geology of Petroleum and Salt . . . , by T.S. Hunt," GSC, *Report of Progress, 1863-66* (1867).

   Hunt suggested that parties search only in pronounced anticlinals and added futher that petroleum would be retained in these only if they were covered by an impervious stratum – such as clay or shales. Otherwise, the petroleum would already have escaped to the surface or saturated higher porous strata. Bell offers as an example of the latter, the recently discovered oil-saturated sands in Athabasca (Bell, "The

Petroleum Fields of Ontario," pp. 102-4). From the earliest days of the petroleum industry, people attempted to solve the technological problem of profitably extracting petroleum from oil-soaked sands and shales – a problem that is still being worked on today. See, for example, the patented process of George J. D'Arcy, an Oil Springs dealer in oil and oil sands for extracting gas, lubricating oil, burning oil, and tar from shale rock (patented in Canada, 31 July 1866, Canada Patent No. 2,085). GSC, *Geology of Canada* (1863), pp. 784-85, contains a surprising account of an 1859 works for distilling shale near the town of Collingwood. Local oil-soaked shale was broken into small fragments and heated in cast-iron retorts to produce illuminating and lubricating oils. Nothing further is known of this operation.

16. *Ontario Commission on Mineral Resources,* pp. 156, 159.
17. Ibid.
18. "Oil Wells in Enniskillen," *Journal of the Board of Arts and Manufacturers for Upper Canada* 1 (June 1861):145-46, cited in Ball, "Petroleum Technology," p. 96.
19. "Trench's Pump Apparatus for Superseding Bags in Artesian Wells," 29 September, 1866 (Canada Patent No. 2,124), by Wm. E. Trench, a Clifton [sic], Ontario, railway conductor; "Non-Corroding Oil Well Packer," by Peter Babcock, an Oil Springs oil operator, with Charles O. Fairbank, Petrolia oil operator (son of John Fairbank), 5 October 1887, No. 27,746; and "Oil Well Pump Packer," by James H. Hoskins, Oil Springs, 1 March 1888, No. 28,568.
20. See Eugene David Thoenon, "Petroleum Industry in West Virginia Before 1890," (Ph.D. dissertation, University of West Virginia, 1956), chapter Four, and Ball, "Petroleum Technology," pp. 31-36.
21. The incidence of hand-, horse-, and steam-powered operations in the early years is detailed in Ball, "Petroleum Technology," pp. 41-47.
22. Toronto *Globe,* 12 September 1862; Ibid., 12 March 1862; and *Canadian News,* 6 March 1862, p. 151, all cited in Ball, "Petroleum Technology," pp. 41-47.
23. See *Ontario Commission on Mineral Resources,* pp. 154, 157, 159; and Bell, "The Petroleum Fields of Ontario," p. 111.
24. See Note 2.
25. Phelps, "John Henry Fairbank," p. 30.
26. *Ontario Commission on Mineral Resources,* p. 159.
27. Dianne Newell, "Technological Innovation and Persistence in the Ontario Oilfields: Some Evidence from Industrial Archaeology," *World Archaeology* 15, 2 (1983):184-95.
28. Bell, "The Petroleum Fields of Ontario," pp. 111-12.
29. The latter was seen as a major contributor to the poor showing of Ontario illuminating oils employed in Nova Scotian lighthouses (see Canada, Department of Marine and Fisheries, *Annual Report, 1870* [1871], "Report of the Nova Scotia Branch," p. 23).
30. See "Compound Silicate Barrel Cement" (Canada Patent No. 1,968); "Centrifugal Barrel Cementer" (No. 1,969); "The Paraffin Barrel Cement" (No. 1,970); "The Resistent Gluten Compound" (No. 1,976), all dated 23 February 1866; "Union Cementing Process for Cementing Petroleum and other Barrels and Vessels" (No. 2,060), 13 June 1866; and "Direct Force Cementing Process" (No. 2,110), 11 September 1866.
31. Patented by James O. Woodruffe, 4 July 1865. *Scientific American* n.s. 13, 26 (1865): 399.
32. This problem is discussed in Canada, Department of Marine and Fisheries, *Annual Report, 1870* (1871), p. 32; and John MacMillan to John Fisken, 23 January 1878, John Fisken Papers, Business Correspondence, Series I, 1877-1878, MU 1041, PAO [hereafter John Fisken Papers].
33. The methods are discussed in *Ontario Commission on Mineral Resources.* See, for example, the testimony of a refiner, M.J. Woodward, p. 157; and Ball, "Petroleum Technology," chapter 4. See also Edward Phelps, "Foundations of the Canadian Oil Industry, 1850-1866," in *Profiles of a Province,* pp. 156-65.
34. See, for example, Canadian patents for improvements in the manufacture of oil gas from petroleum granted to Wm. Watson, 23 March 1861, No. 1,216; Henry Lawson, 29 November 1861, No. 1,297; E.B. Shears, 15 September 1862, No. 1,432; and James E. Thompson and Henry Y. Hind, 28 March 1862, No. 1,344. Richard Fuller was granted a patent for a method to preserve wood with petroleum products (6 May 1864, No. 1,695).
35. Thus, John Fairbank employed a Shaw patent still in his first refinery (Phelps,

"John Henry Fairbank," p. 31, who cites Harkness, "Makers of Oil History," p. 17).

36. Patented 16 December, Canada Patent No. 1,308. A full description of Shaw's still appeared in the Toronto *Globe*, 12 March 1862, and is reproduced and cited in Phelps, "John Henry Fairbank," p. 31, and Ball, "Petroleum Technology," pp. 141-42.
37. Patented 4 December 1861, No. 1,304, and 6 June 1862, No. 1,372, respectively.
38. See, for example, "Canadian Rock Oil Deodorizer," patented by E. Stead, of Toronto (26 March 1862, No. 1,343); and "Tindall's Process for Deodorizing Paraffin, Coal, Pitch, Rock, and other Oils and Hydro-Carbons," by John Wm. W. Tindall, chemist, of Sarnia (24 March 1863, No. 1,513).
39. See "Loftus Process for Refining Petroleum," 21 March 1864, No. 1,664.
40. See *Ontario Commission on Mineral Resources*, pp. 157, 164.
41. Cronin, "North America's Father of Oil," p. 20.
42. The best review of his activity is to be found in the annual reports of the Department of Inland Revenue (1868-1890) and of the Department of Marine and Fisheries (1870-1880). See also Robert Page, "The Early History of Imperial Oil 1880-1900: Technology, Markets, Capital, and Continental Integration" (paper read before the Annual Meeting of the Canadian Historical Association, Vancouver, 6-8 June 1983), p. 6. Page cites Imperial Oil Archives, *Letterbook*, 1880-1890.
43. Bell, "Petroleum in Ontario," p. 111. See also several separate descriptions in J.E. Brantly, *History of Oil Well Drilling* (Houston: Gulf Publishing, 1971), pp. 185, 187-88, 512-14.
44. *Ontario Commission on Mineral Resources*, pp. 157-59.
45. "Report of a Committee of the Trinity-House, London, to that Corporation, on the Fog Signals and Light Houses of Canada and the United States," in Canada, Department of Marine and Fisheries, *Annual Report, 1872* (1873), App. 37 [hereafter, "Report of the Trinity-House Committee"], p. 369.
46. Ball, in "Petroleum Technology," p. 64, writes that notice of this trial was given in the celebrated work of Robert Hunt, ed., *Ure's Dictionary of Manufactures and Mines*, 6th ed., vol. 3 (London: Longmans, Green and Co., 1867), p. 405; and reports of the constant breakdown of the system were carried in *Canadian News*, 10 August 1865, p. 8; and 21 September 1865, p. 187 (both articles were reprinted from reports in the *Oil Springs Chronicle*). The Fauvelle System (developed in the late 1840s in France) is thoroughly described in *Transactions of the North of England Institute of Mining Engineers* 2 (1853-54):60.
47. See H.J. Morgan, ed., *The Canadian Men and Women of their Times*, 2d. ed. (Toronto: Wm. Briggs, 1912), p. 659-60; Brown, *Ideas in Exile*, pp. 67-68; and Brantly, *History of Oil Well Drilling*, pp. 185, 187-88, 204, 496-97, and 512-14. On MacGarvey's career in Galicia, Brantly cites the work of Albert Fauck, *Petroleum in Galicia* (1884). See also "Cost of 'Drilling Rig' complete at Petrolea," in "Brumell's Report on Natural Gas and Petroleum," App. B. For the work of Canadian drillers outside the province, see *Ontario Commission on Mineral Resources*, pp. 154, 160, 166. The Enniskillen driller Duncan Sinclair testified that he had drilled in Italy and New Brunswick and that Canadian drillers were going to Australia, where he also planned to drill. Zaslow, in *Reading the Rocks*, p. 116, writes that when the dominion government purchased a diamond drill to examine the prairies for subsurface occurrences of minerals and water, it hired Petrolia engineer John Ward and his assistant, Alexander Macdonald, who drilled at one location in the North West, while a crew supplied by the Fairbank Company of Petrolia drilled at a second. Personal interviews with a number of returned Petrolia drillers formed the basis for Selwyn P. Griffin's article, "Petrolia, the Cradle of Oil-Drillers," *IOR* 14, 4 (1930): 19-24, and another, by John MacLean, "The Town that Rocked the Oil Cradle," *IOR* 39, 3 (1955):6-10.
48. Note 47; and *Canadian Manufacturer* (7 March 1890), p. 160, notice that R.J. Bradly of Petrolia has recently sent a consignment of two carloads of machinery, drilling tools, and other equipment to Sydney, Australia.
49. See "Bringham's Improvement in the Sinking of Well-Tubes" (24 Feburary 1866, No. 1,982) (Bela B. Bringham was a London gentleman); "Gordon's Improved Apparatus for Drilling Oil Wells and Raising Oil" (3 May 1866, No. 2,029), and "Gordon's Improvement in the Operating of the Walking Beam for Oil Wells" (3 May 1866, No.

2,039) (Alexander Gordon was a Hamilton mechanic and shoemaker); "Buchanan's Well-Sinker and Tubular Well" (18 April 1867, No. 2,320); "Hibbard's Improved Tube-Well" (28 June 1867, No. 2,384); "Lancaster's Improvement in Tube Wells" (12 May 1869, No. 3,917); "Powell's Rod Coupling for Securing and Fastening Together Wood or Iron Rods for Pumps, etc." (13 April 1866, No. 2,016); Eakins' Improvement in the Valves used for Pumping Oil Wells" (12 October 1866, No. 2,133); "Draper's Improvements in the Construction of Pump Valves" (29 April 1871, No. 948), by Tronson Draper, a Petrolia machinist; "Foe's Combination Pump Sucker" (9 May 1871, No. 970); John Walker, Petrolia, "Pump for Oil Wells" (1 August 1884, No. 19,878); John J.H. McKenzie, a Petrolia carpenter, "Adjustable Triangle for Pumping Oil Wells" (24 April 1886, No. 23,894); and Daniel Rosford, Oil Springs, "Drilling Jar for Sinking Wells" (6 May 1887, No. 26,620).

50. Ontario, *Statutes,* 35 Vict., 1871-72, c. 83, "An Act to Provide for the Filling up of or Otherwise Shutting Off the Water Flowing into Abandoned Oil Wells." The act enabled injured parties to apply to a municipal council for permission to stop the flow of water. The original owners, who were not obligated to fill them in, were to be notified. The general problem is discussed in *Popular Science Monthly* 8 (June 1875):149; and John Fairbank mentions the nuisance caused by water-filled oil wells in his testimony before the Ontario commission (see *Ontario Commission on Mineral Resources,* p. 159).
51. The most popular device in the early North American oilfields, including the western peninsula of Ontario, was the Roberts Torpedo, developed for the West Virginia oilfields in 1862, patented in 1865. The innovation and litigation surrounding its imitation and use is carefully documented in Thoenon, "Petroleum Industry in West Virginia," pp 106-9. Its successful use in the Titusville, Pennsylvania region was reported in the *Scientific American* n.s. 15, 4 (1866):54, and according to the *Canadian Mining Review* 6, 12 (December 1888):137, it was extensively employed in 1881 to reopen old wells in the southwestern Ontario petroleum region.
52. *Ontario Commission on Mineral Resources,* p. 160.
53. One, by the Kingston physician Otto Rotton, whose barrel-cementing patents are discussed above, in Note 30, employed the principle of forcing oil up by the static pressure of a column of water introduced into the well; the claim was for the induction pipe (30 January 1866, Canada Patent No. 1,947). This device was expressly "to render into reproduction wells believed exhausted, hence abandoned." The other was that of Sarnia oil producer Henry C. Crocker for removing paraffin that could clog wells from the walls of petroleum deposits (5 September 1887, No. 27,578). He proposed that live steam or hot water be injected into wells.
54. Members of the Trinity-House Committee in 1871 visited the "Maggie Well" in Enniskillen where they reported that gas coming from the well was intercepted at the top by a small pipe fitted into clay packing. From there it was conveyed to the furnace to fuel the boiler, on top of which rested the steam pumping engine for the well. Ten feet farther up the pipe, gas was intercepted again to accumulate in a chamber above, creating a vacuum that assisted the flow of oil through the pipes ("Report of the Trinity-House Committee," p. 369).
55. "Means and Apparatus for Generating Hydrocarbon Vapour and Continuing a Uniform Flow," patented 29 September 1866, No. 2,122.
56. *Ontario Commission on Mineral Resources,* p. 165; and Phelps, "John Henry Fairbank," pp. 55-56.
57. J. Fraser and Co., Buffalo, New York, was the United States patent agency who filed in behalf of the patentee, Thomas M. Ottley (3 May 1866, Canada Patent No. 2,030).
58. See Bell, "The Petroleum Fields of Ontario," p. 112; *Ontario Commission on Mineral Resources,* various testimonies pp. 156-57; and Ball, "Petroleum Technology," chapter 5.
59. *Ontario Commission on Mineral Resources,* p. 165 (Noble's testimony).
60. "Report of the Trinity-House Committee," p. 367. Another response was that of Stephen Webster of Hamilton, who in 1875 patented a metal oil tank to be submerged in water (Patented 25 September 1875, Canada Patent No. 5,195 [extended 22 Sept. 1880, No. 11,797]).
61. *Ontario Commission on Mineral Resources,* pp. 166; "Report of the Trinity House Committee," pp. 367-69; and Phelps, "John Henry

Fairbank," chapter 3.

62. Arthur Menzies Johnson, "The Development of American Petroleum Pipe Lines: A Study in Enterprise and Public Policy, 1862-1906" (Ph.D. dissertation, Vanderbilt University, 1954).
63. *Ontario Commission on Mineral Resources*, pp. 163-66.
64. Canada, *Statutes*, 31 Vict., pt. 2, 1868, c. 50, "An Act to ... Impose an Excise Duty on Refined Petroleum, and to Provide for the Inspection thereof"; Ibid., 40 Vict., 1877, c. 12 (repeal of the excise duty on refined petroleum); Ibid., c. 14 (continuation of the inspection of refined petroleum); and Ibid., 43 Vict., 1886, c. 21, "An Act to Amend the Act Respecting the Inspection of Petroleum."
65. Ibid., 31 Vict., 1868, c. 50, sect. 11.
66. Ibid., 40 Vict., 1877, c. 14, Sect. 4. The fire-test, for example, was lowered to 105 degrees Fahrenheit. It was now permissible to manufacture petroleum below even that standard, but the package in which such petroleum was contained had to be branded accordingly. Nothing of the manner and apparatus for testing was spelled out in the statute, for the department had not yet settled this crucial issue.
67. Ibid., 43 Vict., c. 21, sects. 1-4. Petroleum was now defined as having a specific gravity of not less than 7.75 pounds per gallon. Products with a lower specific gravity were classified as "naphtha". The flash test for domestic refined petroleum was re-established at 115 degrees Fahrenheit. If vapour flashed at a lower temperature or if the gravity of petroleum was heavier than 8.02 pounds per gallon, the material could not be sold for use in Canada for illuminating purposes. Naphtha, however, could be sold in Canada for illuminating purposes, but only to the following conditions: in street lamps, where only the vapour was burned; in dwellings, factories, and businesses when vapourized in secure underground tanks located outside the buildings in question; and mechanical or chemical use in non-residential dwellings.
68. *Ontario Commission on Mineral Resources*, pp. 166, 210.
69. "Report of the Commissioner of Inland Revenue, 1880," in Canada, *Sessional Papers*, 1881, No. 4, p. xxii. As was the case with the other inspection acts, however, this one of 1880 also remained silent as to what standard instrument, if any, should be used to determine the flash test. As the flash test involved many elements of uncertainty, the department made detailed investigations of practices in other countries and conducted controlled experiments in its own laboratories using various pyrometers and thermometers.

    The 1880 annual report of the Commissioner of Inland Revenue provides results of this investigation. The Tagliabues Pyrometer used by inspectors since 1868 to test Canadian petroleum products was found unsatisfactory in that it permitted, and even invited, serious errors. But while the department's investigators were impressed with the English standard pyrometer, they rejected its adoption in Canada because it was, in their words, "so positive and searching in its action that the earliest formation of vapor is detected." If such an instrument were used in Canada, the flash test would effectively have been raised to 125 degrees Fahrenheit, a standard that was deemed too high for economic reasons. The department decided, instead, simply to improve upon the Tagliabues Pyrometer by incorporating elements of the English instrument.
70. See Canada, Department of Inland Revenue, *Annual Report*, 1870 to 1873; Canada, Department of Marine and Fisheries, *Annual Report*, 1872 (1873), pp. 31-33; and "Report of the Trinity-House Committee,' p. 351.
71. "Report of the Trinity-House Committee," p. 357. Increasingly, oil was being stored in metal containers. See Canada Patent No. 19,637, 21 June 1884, for a metallic oil barrel, patented by James W. Cuthbertson and James D. Anderson, both of Bothwell.
72. *Ontario Commission on Mineral Resources*, p. 157.
73. "Report of the Trinity-House Committee," p. 368; and *Ontario Commission on Mineral Resources*, pp. 157-58, 161, and 162-66. Patents for improvements in the manufacture of illuminating gas from petroleum were granted to John Johnson, 11 December 1866, No. 2,180; Alexander G. Alexander, 29 September 1866, No. 2,122; Thomas Alexander, 7 January 1866, No. 2,193; George D'Arcy, 31 July 1868, No. 2,685; Wm. Muir, 23 May 1867, No. 2,350; James A. Grant and James Perry, 1 August 1868, No. 2,690; G. W. Waterhouse, 4 June 1868,

No. 2,618; and N. and H. Totten, 4 May 1870, No. 394. None of the patentees was in the oil business – though one (Perry) was a gas works superintendent. A patent was granted to the London oil operator Samuel Vivian for improvements in the manufacture of machine oil or grease from petroleum or tar (23 May 1871, No. 996). As well, six separate methods were patented for apparatus and for burning petroleum as fuel, and these were granted between 1866 and 1875. See Robert Loudon, 3 October 1866, No. 2,127; O.T. Bevan, 13 March 1867, No. 2,273; Frederick Cook (an Oil Springs refiner) five patents between 1 April 1867 and 3 February 1868 (Nos. 2,294, 2,337, 2,465, 2,477, and 2,479); James Howard, 22 January 1868, No. 2,434; J. Macdonald (Petrolia boilermaker), 11 September 1869, No. 48; and John Parson, Joseph Barret (both Petrolia oil operators), and Robert Marwick, 23 May 1873, No. 2,390. Joseph Beaumont, a Petrolia doctor, patented an artificial coal made of petroleum tar and straw or sawdust (5 April 1875, Canada Patent No. 4,565).

74. *Ontario Commission on Mineral Resources,* p. 155, 157.
75. Ibid., p. 163. See also Ball, "Petroleum Technology," chapter 3, "Changing a Marketable Product." William Spencer is another who has been credited with introducing the lead process (Maurice Gillis, "He Took the 'Skunk' out of Oil," *IOR* 40, 3 [1956]:8-10).
76. See 27 June 1866, No. 2,054; 10 April 1867, No. 2,303; 6 June 1867, No. 2,360; 9 March 1867, No. 2,259; 17 October 1869, No. 96; 27 July 1870, No. 524; 2 October 1876, No. 6,631; 9 November 1876, No. 6,746; 28 September 1876, No. 6,580; 26 July 1879, No. 10,299; 28 May 1879, No. 10,018; 8 September 1880, No. 11,726; and 10 May 1881, No. 12,770; 18 June 1883, No. 17,016; 25 October 1886, No. 25,201; and Notes 77 and 81.
77. "Kittridge's Process of Refining Petroleum," 7 November 1886, Canada Patent No. 25,089; "Woodward's Process of Refining Petroleum and Other Substances containing Sulphur or Phosphorus," 30 November 1886, No. 25,301; and "Woodward's Method and Apparatus for Increasing the Vapour Test of, and Partly Purifying, Petroleum Distillate," 31 December 1886, No. 25,664. See Kittridge's testimony before the *Ontario Commission on Mineral Resources,* pp. 161-62.
78. Canada, Department of Marine and Fisheries, *Annual Report, 1872* (1873), pp. 358, 369.
79. 10 May 1881, Canada Patent No. 12,770.
80. *Ontario Commission on Mineral Resources,* p. 161. It is to be assumed that the recovered sulphuric acid was not suitable for reuse in refining operations.
81. See biographies of Herman Frasch in *Dictionary of American Biography,* vol. 6 (New York: Scribner's, 1931), pp. 602-3; and *National Cyclopaedia of American Biography,* vol. 19 (New York: James T. White, 1926), pp. 347-48. A fairly detailed but undocumented account of Frasch's London-based activities is to be found in Gillis, "He Took the 'Skunk' out of Oil," pp. 8-10. The Frasch process was patented in Canada as follows: 24 April 1884, No. 19,189 (assigned to Imperial Oil); 11 May 1886, No. 24,034; 23 June 1887, No. 27,033; and 23 March 1888, No. 28,750. He is responsible for an additional nine patents (issued from an Ohio address) in 1891.
82. See Page, "The Early History of Imperial Oil;" and John T. Saywell, "The Early History of Canadian Oil Companies: A Chapter in Canadian Business History," *Ontario History* 53 (1961):67-72.
83. 20 November 1874, Canada Patent No. 4,079; and 27 February 1875, No. 4,443.
84. The formation of The Petroleum Refining Company of Canada, an association of refiners, is reported in Canada, Department of Marine and Fisheries, *Annual Report, 1872* (1873), p. 32. See also Phelps, "John Henry Fairbank," chapter 2, especially pp. 32-36 and 65-70; Phelps, "Foundations of the Canadian Oil Industry," pp. 161-64; and E. Phelps, "The Canadian Oil Association – An Early Business Combination," *Western Ontario Historical Notes* 19 (September 1963):31-39.
85. Phelps, "John Henry Fairbank," pp. 56-62. Phelps attributes the impermanence of these organizations to their loose and voluntary nature (p. 62). Sally Zerker's study of centralization in the oil industry looks at the economic reasons for monopolistic behavior in the oil industry ("Centralization in the International Oil Industry" [paper read before the Annual Meeting of the Canadian Historical Association, Vancouver, 6-8 June 1983]).

86. John Fisken Papers, John MacMillan to John Fisken, 23 January 1878; R. Stobe to John Fisken, 8 November 1877; Star Oil Company to John Fisken, 6 October 1878; Home Oil Works to John Fisken, 28 January 1878; and Mutual Oil Refining Company to John Fisken, 22 November 1878. John Fisken and Co. were Toronto produce and commission merchants who reincorporated in 1864 to serve the oil business. Refiners in the province supplied Fisken, who then marketed to individual retailers. The business correspondence concerns the oil business for 1878, especially as it relates to shipment problems and oil prices.
87. GSC, *Mineral Statistics 1887* (1888), p. 57. See also F. Cronin, "They Called it London-in-the-Bush," *IOR* 39, 4 (1955):21-24, for a brief sketch; and Dianne Newell, "'All in a Day's Work': Local Invention on the Ontario Mining Frontier," *Technology and Culture* (forthcoming October 1985).
88. See *Canadian Mining Review* 8 (1889):7; RG 87, Department of Mines, Mineral Statistics, vol. 27, file 116, PAC; and MacLean, "The Town that Rocked the Oil Cradle," pp. 17-20. For information on Imperial's president, Jake Englehart, see Ian Sclanders, "The Amazing Jake Englehart," *IOR* 39, 4 (1955):2-7.
89. I am indebted to Morris Zaslow for the suggestion that the attempts by Imperial Oil to monopolize the Ontario oil business simply made the eventual takeover smoother.
90. See various reports by T.S. Hunt published in the Geological Survey, *Report of Progress,* especially his report for 1863-66, published in 1867; and *Ontario Commission on Mineral Resources,* p. 190.
91. "On the Goderich Salt Region . . . , by T.S. Hunt," in GSC, *Report of Progress, 1866-69* (1870) [hereafter, "Hunt's Report on the Goderich Salt Region for 1866-69"], pp. 222-23.
92. "Observations on the History and Statistics of the Trade and Manufacture of Canadian Salt, by J.L. Smith," in GSC, *Report of Progress, 1874-75* (1876) [hereafter, "Smith's Report on Salt"], pp. 268, 287.
93. Ibid., p. 284.
94. "Hunt's Report on the Goderich Salt Region for 1866-69."
95. "The Goderich Salt Region and Mr. Attrill's Exploration, by T.S. Hunt," in GSC, *Report of Progress, 1876-1877* (1878); and T.S. Hunt, "The Goderich Salt Region," *Transactions of the American Institute of Mining Engineers* 5 (1876-77):538-60.
96. "Hunt's Report on the Goderich Salt Region for 1866-69."
97. Ibid., pp. 182-83.
98. Ibid., p. 190. It was the Excelsior Salt Works (see Department of Mines, Mineral Statistics, vol. 30, file 127, PAC).
99. For a description of the Onondaga system, see *Engineering* (8 August 1887):103-4. T.S. Hunt's report on salt for 1866-69 enumerates the use of kettles (the Onondaga system) in the first years. The Dominion Salt Works (Goderich) had 60; the Goderich Salt Company, the second largest producer in 1868, 104 kettles; Victoria Salt Company (Goderich), 60 kettles; Ontario Salt Company (Goderich), also 60 kettles; and Huron Salt Works (Goderich), the latest in the district in 1868, 120 large kettles arranged on the pattern of the Onondaga Company Works (see "Hunt's Report on the Goderich Salt Region for 1866-69," pp. 212-42). Smith's 1874 examination of the Ontario salt district lists each of the works, the number of blocks, pans, horsepower of engines, and the system of evaporation ("Smith's History of Salt," pp. 294-95).
100. "Hunt's Report on the Goderich Salt Region for 1866-69," p. 223.
101. Ibid., pp. 224-25, and "Smith's History of Salt," p. 295; *Engineering* (8 August 1879):103.
102. "Hunt's Report on the Goderich Salt Region for 1866-69," p. 223-24, 239.
103. See a description of the Saginaw salt processing in *Scientific American* n.s. 7, 13 (1862):193. The newly patented Garrison Salt Block employed a complex and elaborate arrangement of pans heated by steam or hot air circulated under and around the pans. See also that of Nathan Chapin (ibid., 4, 7 [1862]:97; and Ibid., 11, 17 [1864]:257). The Chapin Salt Block used tubular furnaces that ran the full length of the pans.
104. "Hunt's Report on the Goderich Salt Region for 1866-69," p. 225-26.
105. Ibid.
106. "Smith's History of Salt," table 3, pp. 294-95.
107. Ransford did, however, patent a brine car, whereby wet salt taken from the pans was further drained before treatment (26 May 1871, No. 1,001).
108. "Platt's Salt Evaporator," patented 17 July

1868, No. 2,686. See "Hunt's Report on the Goderich Salt Region for 1866-69," pp. 225-26. Two Perth, Ontario, men, W.J. Morris, the bolt manufacturer, and T. Aspden, a chemist, patented a salt block (3 September 1869, No. 3,314) (see also, 25 September 1868, No. 2,798, and 18 April 1871, No. 888), but it is not known if it was ever adopted in the salt district.

109. "Smith's History of Salt," table 3, pp. 294-95; 279 (23 October 1869, No. 3,315).
110. See Ibid.
111. *Ontario Commission on Mineral Resources,* p. 188. Herbert Harrison, another Goderich salt manufacturer, and N. Stickney from Dracut, Mass., patented a verticle brine cylinder with encircling steam chamber, 25 September 1876, No. 6,574.
112. See, for example, the process for removing gypsum or lime from pans patented by a Toronto chemist, Robert W. Elliot, 2 February 1871, No. 811.
113. Martin Peter Hayes, a Seaforth banker. The patent is listed as 16 June 1871, Canada Patent No. 1,030. See also his further improvements in the patents issued 12 February 1873, Canada Patent No. 2,052; 23 May 1873, No. 2,401; (with P. McEwan) 11 June 1873, no. 2,431; and (with Thos. B. Wilson, Manchester, England), 13 August 1875, No. 5,058. See, also, "Smith's History of Salt," table 3, pp. 294-95.
114. Ibid., 10 June 1871, Canada Patent No. 1,505.
115. Ibid. Interestingly, John Ransford owned large tracts of woodlands, so he was a seller of cordwood, not a purchaser ("Smith's History of Salt," p. 300).
116. Ibid., p. 283. Smith noted that similar contemporary experiments took place in the United States that produced identical conclusions.
117. See *Ontario Commission on Mineral Resources,* pp. 189, 191. Slack coal was used at the Ontario Peoples Salt Manufacturing Company, a Kincardine farmers' co-operative.
118. Ibid., p. 191, testimony of James Carter, secretary-treasurer of the Courtright Salt Company. This company was established in 1884 and began using petroleum tar for fuel in 1885. Carter himself had patented an invention, "Side Return Flue Salt Evaporator," 12 April 1873, No. 2,254.
119. See "Foster's Waste-Steam Utilizer," patented 17 July 1868, No. 2,691. Charles Foster was a Middlesex County farmer.
120. "Smith's History of Salt," pp. 276-77.
121. Ibid., p. 277; and *Ontario Commission on Mineral Resources,* p. 190-91. A version of the system of using wooden vats and metal heating tubes was patented by an Egmondville merchant, Geo. E. Jackson, 28 August 1888, No. 29,755.
122. *Ontario Commission on Mineral Resources,* p. 188ff.
123. Ibid.
124. Ibid., pp. 184, 191.
125. This I infer from the information contained in Canada Department of Mines, Mineral Statistics, vol. 27, file 116, PAC.
126. Ibid.

BIBLIOGRAPHY

PUBLISHED WORKS

A. Government Publications

Ontario provided few reports on provincial mining prior to 1891, when the Ontario Bureau of Mines was established, though the annual reports of the provincial department of Crown lands are quite helpful. The report of the Royal Commission on the Mineral Resources of Ontario, 1890, however, represents a singularly outstanding historical source without which this present study would have been impossible. It was the regular reports of the Geological Survey of Canada that furnished the bulk of the information on provincial mining. The specific mentions are too numerous to list here, but interested readers can consult the notes in the case study chapters 4 through 6. The most important and detailed ones next to William Logan's early reports on the pioneer operations of the Upper Great Lakes are those of T.S. Hunt on early petroleum and salt developments in the Western Peninsula district and Henry Vennor's excellent mining studies, published annually from 1869 to 1880, on various mining operations in the Southeastern district. After 1884 the Survey established a mines section and appointed mining engineers like Brummel, Coste, and Ingall, who were charged with producing lengthy monographs, with historical accounts and details of technology employed, on all three Ontario mining districts. Lastly, Rossiter Raymond's studies of mining methods and mining education in the U.S. provide excellent comparative data for Canada.

Andrews, P.W. "The Canadian Mineral Industry and the Economy." Canada Department of Energy, Mines and Resources, Mineral Resources Branch, *Mineral Bulletin* MR128 (1972).

Bishop, Olga B. *Publications of the Government of Ontario, 1867-1900.* Toronto: Ministry of Government Services, 1976.

———. *Publications of the Government of the Province of Canada, 1841-1867.* Ottawa: National Library, 1963.

Blue, Archibald. "The Story of Silver Islet." Ontario Bureau of Mines, *Report, 1892* (1892), pp. 125-57.

Bush, Edward Forbes. "Commercial Navigation on the Rideau Canal, 1832-1961." *History and Archaeology* 54, Ottawa: Parks Canada, 1981.

Canada. *Census of Canada, 1870-71.* Vol. 2. Ottawa: I.B. Taylor, 1873, Tables 13 and 27.

———. *Census of Canada, 1880-81.* Vol. 2. Ottawa: MacLean, Roger and Co., 1884, Tables 14, 28, and 44.

———. *Census of Canada, 1890-91.* Vol. 2. Ottawa: S.E. Dawson, 1893, Table 12.

———. Department of Indian Affairs. *Annual Report.* "Reports of Superintendent and Agents," 1880-1890.

———. Department of Inland Revenue. *Annual Report.* "Reports, Returns and Statistics of the Inland Revenues of the Dominion of Canada," 1870-1890.

———. Department of Marine and Fisheries. *Annual Report,* 1870-1880.

———. Department of Trade and Commerce, Dominion Bureau of Statistics. *Chronological Record of Canadian Mining Events from 1604 to 1947 and Historical Tables of Mineral Production in Canada.* Ottawa: King's Printer, 1948.

———. Geological Survey of Canada. *Report of Progress,* 1866-69 to 1879-80.

———. ———. *Annual Report,* 1884-1891.

———. National Research Council of Canada. Canada Institute for Scientific and Technical Information. *Union List of Scientific Serials in Canadian Libraries.* 2 Vols. 6th ed. Ottawa: Queen's Printer, 1975.

———. Parliament. *Statutes,* 1868-1890.

———. Patent Office. *List of Canadian Patents from the Beginning of the Patent Office, June, 1824, to the 31st of August, 1872.* Ottawa: MacLean, Roger and Co., 1882.

Canada (Province). *Census of the Canadas, 1851-52. Personal Census,* Vol. 1. Quebec: John Lovell, 1853, App. 7.

———. *Census of the Canadas, 1860-61. Personal Census,* Vol. 1. Quebec: S.B. Foote, 1863, App. 8 and 11.

———. Department of Crown Lands. *Report of the Commissioner of Crown Lands,* 1856-1866.

———. Geological Survey of Canada. *Report of Progress,* 1844 to 1863-66.

———. ———. *Report of Progress from its Commencement to 1863.* Montreal: Dawson Bros., 1863.

———. Legislative Assembly. *Appendix to the Journals,* 1844/45 to 1859.

———. ———. *Sessional Papers,* 1860-1867.

———. *Statutes,* 1841-1866.

Coleman, Margaret. "The Canadian Patent Office from the Beginning to 1900." National Historic Parks and Sites Branch [Canada], *Research Bulletin* 32 (1976).

Convey, John; and Rabbits, F.T. "The Dissemination of Technical Information to Canadian Industry." Canada Department of Mines and Technical Services, Mines Branch, IC165 (1964).

Fleming, Sanford. *Report on Surveys and Preliminary Operations on the Canadian Pacific Railway up to January, 1877.* Ottawa: Queen's Printer, 1877.

Gibson, Thomas W. *Mining in Ontario.* Toronto: King's Printer, 1937.

Great Britain. Royal Commission on Technical Instruction. *Report on Technical Education in the United States of America and Canada,* by William Mather (1881). Reprinted in *Irish University Press Series of British Parliamentary Papers.* Shannon: Irish University Press, 1970, Vol. 6.

Hewitt, D.F. "Salt in Ontario." *Industrial Minerals Report No. 6.* Toronto: Ontario Department of Mines, 1962.

Johnston, A.G., comp. *Index of Publications, Geological Survey of Canada (1845-1958).* Ottawa: Department of Mines and Technical Surveys, 1961.

Malholtra, S.P. "The Development of Canadian Mineral Statistics and Confidentiality under Federal and Provincial Statistical Acts." Canada Department of Energy, Mines and Resources, Mineral Resources Branch, MR136 (1973).

Ontario. Bureau of Mines. *Report,* 1891-1895.

———. Department of Crown Lands. *Report of the Commissioner of Crown Lands of the Province of Ontario,* 1867-1890.

———. Division of Mines. "Ontario Mineral Map." Map 2310. 1974.

———. Legislative Assembly. *Sessional Papers,* 1868/69 to 1890.

———. ———. *Statutes,* 1867/68 to 1890.

———. *Report of the Royal Ontario Nickel Commission.* Toronto: King's Printer, 1917.

———. *Royal Commission on the Mineral Resources of Ontario and Measures for their Development.* Toronto: Warwick and Sons, 1890.

Palmer, Walter William. "A Pioneer's Mining Experience on Lake Superior and Lake Huron." Ontario Bureau of Mines, *Report, 1892* (1892), pp. 171-78.

Peckham, S.F. "Production Technology, and Tests of Petroleum and Its Products." In *Tenth Census of the United States.* Washington, D.C.: U.S.G.P.O., 1884.

Stacey, Duncan A. *Sockeye and Tinplate: Technological Change in the Fraser River Canning Industry 1871-1912.* Heritage Record no. 15. Victoria, B.C.: Provincial Museum, 1982.

United States. Census Office. *Tenth Census,* 1880. 22 vols. Washington: U.S.G.P.O., 1883-88.

———. 40th Congress. 3d Session. 1869. *Report of Rossiter W. Raymond on the Mineral Resources of the States and Territories West of the Rocky Mountains.* House Exec. Doc. 54, Sect. 7 ["Mining Education," pp. 224-50].

———. 41st Congress. 2d Session. 1870. *Report of Rossiter W. Raymond on the Statistics of Mines and Mining in the States and Territories West of the Rocky Mountains.* House Exec. Doc. 207, sect. 4 ["The Mechanical Appliances of Mining," pp. 471-726; "Metallurgical Processes," pp. 727-759].

———. House of Representatives. Subcommittee on Natural Resources. *Technological Trends and National Policy, Including the Social Implications of New Inventions.* Washington, D.C.: U.S.G.P.O., 1937.

B. Books

Agricola, Georgius. *De re metallica.* Trans. and ed. by Herbert Hoover and Lou Clark Hoover. New York, 1950.

Arthur, Elizabeth D., ed. *The Thunder Bay District, 1821-1892. A Collection of Documents.* Champlain Society of Ontario, Ser. 9. Toronto: University of Toronto Press, 1973.

Berger, Carl. *The Writing of Canadian History. Aspects of English Canadian Historical Writing: 1900 to 1970.* Toronto: Oxford University Press, 1976.

Blainey, G. *The Rush That Never Ended: A History of Australian Mining.* Melbourne: Melbourne University Press, 1963.

Bolton, Henry Carrington. *A Catalogue of Scientific and Technical Periodicals (1665-1882).* Washington: Smithsonian Institution, 1885.

Brantly, J.E. *History of Oil Well Drilling.* Houston: Gulf Publishing, 1971.

Brown, J.J. *Ideas in Exile. A History of Canadian Invention.* Toronto: McClelland and Stewart, 1967.

Brown, Lawrence A. *Innovation Diffusion: A New Perspective.* New York: Methuen, 1981.

Centennial Seminar on the History of the Petroleum Industry. *Oil's First Century.* Cambridge, Mass.: Harvard University Press, 1959.

Chandler, Alfred D., Jr. *The Visible Hand: The Managerial Revolution in American Business.* Cambridge, Mass.: Belknap Press, 1977.

Church, John Adams. *Mining Schools in the United States.* New York: Waldron & Payne, 1871.

Clement, Wallace. *Hardrock Mining: Industrial Relations and Technological Changes at Inco.* Toronto: McClelland and Stewart, 1981.

David, Paul A. *Technical Choice, Innovation and Economic Growth. Essays on American and British Experience in the Nineteenth Century.* Cambridge: Cambridge University Press, 1975.

Dean, W.G., ed. *Economic Atlas of Ontario.* Toronto: University of Toronto Press, 1969.

Donald, W.J.A. *The Canadian Iron and Steel Industry: A Study in the Economic History of a Protected Industry.* Boston: Houghton Mifflin, 1915.

Drinker, Henry Sturgess. *Tunneling, Explosive Compounds and Rock Drills.* New York: Wiley, 1878.

Faucher, Albert. *Québec en Amérique au XIXe siècle: essai sur les charactères économiques à la Laurentie.* Montréal: Fides, 1973.

Ferguson, Eugene S. *Bibliography of the History of Technology.* Cambridge, Mass.: M.I.T. and the Society for the History of Technology, 1968.

Firestone, O.J. *Canada's Economic Development 1867-1953 with Special Reference to Changes in the Country's National Product and National Wealth.* International Association for Research in Income & Wealth, Ser. VII. London: Bowes and Bowes, 1958.

Fivehouse, Dan. *The Diamond Drill Industry.* Saanichton, B.C.: Hancock House Publishers, 1976.

Fuller, G.N. *Geological Reports of Douglass Houghton, First State Geologist of Michigan, 1837-1845.* Lansing, Mich., 1941.

Gates, W.B., Jr. *Michigan Copper and Boston Dollars: An Economic History of the Michigan Copper Mining Industry.* Cambridge, Mass.: Harvard University Press, 1951.

Glazebrook, G.P. de T. *A History of Transportation in Canada.* 2d ed. 2 vols. Toronto: McClelland and Stewart, 1964.

Grant, George. *Technology and Empire: Perspectives on North America.* Toronto: Anansi, 1969.

Guillet, Edwin C. *Pioneer Travel in Upper Canada.* 1933. Reprint. Toronto: University of Toronto Press, 1963.

Habakkuk, H.J. *American and British Technology in the Nineteenth Century: The Search for Labour-Saving Inventions.* Cambridge: At the University Press, 1962.

Hägerstrand, Torsten. *Innovation Diffusion as a Spatial Process.* Postscript and Trans. by Allan Pred. Chicago: University of Chicago Press, 1967.

Henderson, E. Marion. *The McKellar Story. McKellar Pioneers in Lake Superior's Mineral Country 1839 to 1929.* Thunder Bay: The McKellar Story Publication Committee, 1981.

Herfindahl, O.C. *Copper Costs and Prices: 1870-1957.* Baltimore: Johns Hopkins University Press, 1959.

Hindy, Ralph W.; and Hindy, Muriel E. *Pioneering in Big Business 1882-1911.* New York: Harper, 1955.

Hodgetts, J.E. *Pioneer Public Service: An Administrative History of the United Canadas, 1841-1867.* Canadian Government Series No. 7. Toronto: University of Toronto Press, 1955.

Hunter, Louis C. *A History of Industrial Power in the United States, 1780-1930.* Vol. 1. *Water Power in the Century of the Steam Engine.* Charlottesville: University Press of Virginia, 1979.

Hyde, Charles K. *Technological Change and the British Iron Industry, 1700-1870.* Princeton: Princeton University Press, 1977.

Innis, Harold A. *Settlement and the Mining Frontier* (Lower, A.R.M. *Settlement and the Forest Frontier of Eastern Canada*). Canadian Frontiers of Settlement, edited by W.A. MacKintosh and W.G.G. Joerg, vol. 9. 1936. Reprint. Millwood, New York: Kraus Reprint, 1974.

Jarrell, R.A.; and Ball, R.N., eds. *Science, Technology, and Canadian History.* The First Conference on the Study of the History of Canadian Science and Technology. Waterloo: Wilfrid Laurier University Press, 1980.

Johnson, A.B. *Recollection of Oil Drilling at Oil Springs, Ontario, with Notes of the Shaw Gusher, the Drake Well in Pennsylvania, the First Flowing Well, and a Description of the Different Operating Systems to the Present Time.* Tillsonburg, Ont.: Harvey F. Johnson, 1938.

Kenwood, A.G.; and Lougheed, A.L. *Technological Diffusion and Industrialization before 1914.* London: Croom Helm, 1982.

Kerr, D.G.G. *Historical Atlas of Canada.* 3d rev. ed. Don Mills, Ont.: Nelson, 1975.

Koether, George. *The Building of Men, Machines and a Company.* Ingersoll-Rand Company, 1971.

Kransberg, Melvin; and Pursell, Carroll W., Jr., eds. *Technology in Western Civilization.* 2 vols. New York: Oxford University Press, 1967.

———; and Davenport, William H. *Technology and Culture: An Anthology.* New York: New American Library, 1972.

Kunhart, W.B. *The Art of Ore Dressing in Europe.* New York: Wiley, 1884.

Lake, James. A. *Law and Mineral Wealth: The Legal Profile of the Wisconsin Mining Industry.* Madison: University of Wisconsin Press, 1962.

Langford, B.B. *Out of the Earth. The Mineral Industry in Canada.* Toronto: University of Toronto Press, 1954.

Le Bourdais, D.M. *Metals and Men: The Story of Canadian Mining.* Toronto: McClelland and Stewart, 1957.

———. *Sudbury Basin: The Story of Nickel.* Toronto: Ryerson, 1953.

Levere, Trevor H.; and Jarrell, Richard A., eds. *A Curious Field-Book: Science and Society in Canadian History.* Toronto: Oxford University Press, 1974.

Longo, Roy M., ed. *Historical Highlights of Canadian Mining.* Toronto: Pitt Publishing, 1973.

Lower, A.R.M. *Settlement and the Forest Frontier in Eastern Canada* (Innis, Harold A., *Settlement and the Mining Frontier*). Canadian Frontiers of Settlement, edited by W.A. MacKintosh and W.L.G. Joerg, vol. 9. 1936. Reprint. Millwood, N.Y.: Kraus Reprint, 1974.

Lower, Arthur R.M. *Great Britain's Woodyard: British America and the Timber Trade, 1763-1867.* Montreal: McGill-Queen's University Press, 1973.

Lucas, Rex A. *Minetown, Milltown, Railtown: Life in Canadian Communities of Single Industry.* Toronto: University of Toronto Press, 1971.

Marr, William M.; and Paterson, Donald G. *Canada: An Economic History.* Toronto: Macmillan, 1980.

Mathews, William, comp. *Canadian Diaries and Autobiographies.* Berkeley: University of California Press, 1950.

McCharles, Æneas. *Bemocked of Destiny.* Toronto: William Briggs, 1908.

McGill University. *Graduates of McGill University, Montreal, Corrected to December, 1901.* Montreal: Witness Printing, 1901.

McGivern, James. *First Hundred Years of Engineering Education in the United States.* Washington: Gonzaga University Press, 1960.

McHugh, Jeanne. *Alexander Holley and the Makers of Steel.* Baltimore: Johns Hopkins University Press, 1980.

Moore, Elwood S. *The Mineral Resources of Canada.* Toronto: Ryerson Press, 1933.

———. *American Influence in Canadian Mining.* Political Economy Series No. 9. Toronto: University of Toronto Press, 1941.

Morgan, H.J., ed. *The Canadian Men and Women of Their Time: A Handbook of Canadian Biography of Living Characters.* 2d ed. Toronto: William Briggs, 1912.

Morris, Hon. Alexander. *The Treaties of Canada with the Indians.* 1880. Reprint. Coles Canadiana Collection, 1971.

National Bureau of Economic Research. *The Rate and Direction of Inventive Activity: Economic and Social Factors.* No. 13. Princeton: Princeton University Press, 1962.

Naylor, Tom. *The History of Canadian Business, 1867-1914.* 2 vols. Toronto: James Lorimer, 1975.

Nelles, H.V. *The Politics of Development: Forest, Mines, and Hydro-Electric Power in Ontario, 1849-1941.* Toronto: Macmillan, 1975.

Nevins, Allan. *John D. Rockefeller: The Heroic Age of American Enterprise.* Vol. 1. New York: Scribner's, 1940.

Noble, David F. *America By Design: Science, Technology, and the Rise of Corporate Capitalism.* New York: Knopf, 1977.

North, Douglass C.; and Thomas, Robert Paul. *The Rise of the Western World: A New Economic History.* Cambridge: At the University Press, 1973.

Parsons, A.B., ed. *Seventy-Five Years of Progress in the Mineral Industry 1871-1946.* New York: AIME, 1947.

Paul, Rodman W. *Mining Frontiers of the Far West, 1848-1880.* Toronto: Holt, Rinehart and Winston, 1963.

Peters, Edward D., Jr. *Modern American Methods of Copper Smelting.* New York: Scientific Publishing, 1887.

Pred, Allan R. *Spatial Dynamics of U.S. Urban-Industrial Growth, 1800-1914: Interpretive and Theoretical Essays.* Cambridge, Mass.: M.I.T. Press, 1966.

———. *Urban Growth and the Circulation of Information: The United States System of Cities, 1790-1840.* Cambridge, Mass.: Harvard University Press, 1973.

Pursell, Carroll W., Jr. *Early Stationary Steam Engines in America: A Study in the Migration of a Technology.* Washington: Smithsonian Institution Press, 1969.

Read, Thomas Thorton. *The Development of Mineral Industry Education in the United States.* New York: AIME, 1941.

Richardson, W. George. *A Survey of Canadian Mining History.* Montreal: Canadian Institute of Mining and Metallurgy, 1974.

Rickard, T.A. *The Copper Mines of Lake Superior.* New York: Engineering and Mining Journal, 1905.

Robinson, J. Lewis. *Resources of the Canadian Shield.* Toronto: Methuen, 1969.

Rowe, John. *Cornwall in the Age of the Industrial Revolution.* Liverpool: The University Press, 1972.

———. *The Hard Rock Men: Cornish Immigrants and the North America Mining Frontier.* Liverpool: The University Press, 1974.

Saul, S.B., ed. *Technological Change: The United States and Britain in the Nineteenth Century.* Debates in Economic History, Peter Mathias, ed. London: Methuen, 1970.

Schumpeter, Joseph, *The Theory of Economic Development.* Cambridge, Mass.: Harvard University Press, 1934.

———. *Business Cycles: A Theoretical, Historical, and Statistical Analysis of the Capitalist Process.* New York: McGraw-Hill, 1934.

Scudder, S.H. *Catalogue of Scientific Serials of All Countries Including the Transactions of Learned Societies in Natural, Physical, and Mathematical Sciences, 1633 to 1876.* Cambridge, 1879.

Shortt, Adam; and Doughty, Arthur, eds. *Canada and Its Provinces.* Vol. 18. *Ontario.* Edinburgh: Constable, 1914.

Sinclair, Bruce; Ball, Norman R.; and Peterson, James O., eds. *Let Us Be Honest and Modest. Technology and Society in Canadian History.* Toronto: Oxford University Press, 1974.

Spence, Clark C. *British Investments and the American Mining Frontier, 1860-1891.* Ithaca: Cornell University Press, 1958.

———. *Mining Engineers and the American West: The Lace-Boot Brigade, 1849-1933.* New Haven: Yale University Press, 1970.

Stack, Barbara. *Handbook of Mining and Tunnelling Machinery.* New York: John Wiley, 1982.

Straussman, Paul W. *Risk and Technological Innovation: American Manufacturing Methods During the Nineteenth Century.* Ithaca: Cornell University Press, 1959.

Strickland, Helen Moore. *Silver under the Sea. The Story of the Silver Islet Mine Near Thunder Bay, Ontario, Canada.* Cobalt: Highway Book Shop, 1979.

Tarbell, Ida M. *The History of Standard Oil Company.* New York: Norton, 1969.

Taylor, K.W.; and Mitchell, H. *Statistical Contributions to Canadian Economic History.* Vol. 2. Toronto: Macmillan, 1931.

Torstendal, Rolf. *Dispersion of Engineers in a Transitional Society: Swedish Technicians, 1860-1940.* Upsala: Acta Universistatis Upsaliensis, 1975.

Urquhart, M.C.; and Buckley, K.A.H., *Historical Statistics of Canada.* Toronto: Macmillan, 1965.

Wallace, Anthony F.C. *Rockdale: The Growth of an American Village in the Early Industrial Revolution.* New York: Knopf, 1978.

Wallbridge, Malcolm, ed. *The Diary of Thomas Nightengale, Farmer and Miner, 1867-1871.* Picton, Ont.: Picton Gazette, 1967.

Weiss, John H. *The Making of Technological Man: The Social Origins of French Engineering Education.* Cambridge, Mass.: M.I.T. Press, 1982.

White, Lynn, Jr. *Medieval Technology and Social Change.* Oxford: Clarendon, 1962.

Wiebe, Robert H. *The Search for Order, 1877-1920.* New York, 1967.

Williamson, Harold F.; and Daum, Arnold R. *The American Petroleum Industry.* Vol. 1. *The Age of Illumination, 1859-1899.* Evanston: Northwestern University Press, 1959.

Willimot, Arthur B. *The Mineral Wealth of Canada.* Toronto: William Briggs, 1898.

Young, C.R. *Early Engineering Education at Toronto, 1851-1919.* Toronto: University of Toronto Press, 1958.

Young, Otis E., Jr. *Western Mining.* Norman: University of Oklahoma Press, 1970.

———. *Black Powder and Hand Steel: Miners and Machines on the Old Western Frontier.* Norman: University of Oklahoma Press, 1976.

Zaslow, Morris. *The Opening of the Canadian North, 1870-1914.* The Canadian Centenary Series, No. 16. Toronto: McClelland and Stewart, 1971.

———. *Reading the Rocks: The Story of the Geological Survey of Canada, 1842-1972.* Toronto: Macmillan, 1975.

C. Pamphlets

British and Canadian Mining Company of Lake Superior. *Report of the Trustees of the British and Canadian Mining Company of Lake Superior.* Montreal: Canada Gazette Office, 1847.

British North American Mining Company. *Report of the Directors of the British North American Mining Company.* Montreal: Lovell and Gibson, 1847, 1848.

Canadian Native Oil Company (Ltd.). *The Canadian Native Oil: Its Story, Its Uses, and Its Profits, with Some Account of a Visit to the Oil Wells.* London: Ashky, 1862.

Chapman, E.J. *Report on the Frontenac Lead Mine.* N.p., 1868.

———. *Report of Professor E.J. Chapman on Mineral Locations in North Hastings, Ontario.* Toronto, 1881.

Dawson, J.W. *Report on the Frontenac Lead Mine.* N.p., 1868.

Fraser and Chalmers Company. *Catalogue, No. 3.* Chicago, [1888?].

Gzowski, C.S. *Report . . . Examine and Report Upon the Mines of the Upper Canada Mining Company on Lake Huron.* Hamilton, C.W.: Spectator Office, 1848.

Lake Huron Silver and Copper Company. *Report of the President and Directors of the Lake Huron Silver and Copper Mining Company.* Montreal: Montreal Pilot Office, 1848.

Livingstone, Janey C. *Historic Silver Islet. The Story of a Drowned Mine.* Ft. William: Times-Journal Press, n.d.

McKellar, Peter. *Mining on the North Shore of Lake Superior.* N.p., 1874.

Montreal Mining Company. *Annual Report of the Directors of the Montreal Mining Company to the Stockholders.* Montreal [publisher varies], 1846; 1848-1850; 1853-1854; 1856; 1866; 1869; 1870.

Mowbray, George M. "The Hoosac Tunnel and the Use of Nitro-Glycerine in Blasting." N.p., 1874.

"Prospectus of the Canada Consolidated Gold Mining Company," n.d.

Robb, Charles. *Report on the Frontenac Lead Mine,* N.p., 1868.

Sibley, A.H. *Report on Mining on the North Shore of Lake Superior, by A.H. Sibley, President of the Silver Islet Company.* N.p., 1873.

Silver Islet Consolidated Mining and Lands Company. *Report of the Directors to Stockholders.* N.p., 1879; 1884.

Stewart, Alexander. *Thunder Bay Silver Mining Company Report.* Montreal: J. Starke & Co., 1874.

Upper Canada Mining Company. *Report of the Upper Canada Mining Company.* Hamilton: Spectator Office, 1847, 1848.

D. Contemporary Newspapers and Periodicals

Algoma Miner and Weekly Herald [*Weekly Herald and Lake Superior Mining Journal*], 1882-1890.

American Geologist, 1888-1890.

American Journal of Agriculture and Science, 1845-1848.

The American Minerological Journal (N.Y.), 1814.

British American Magazine (Toronto), 1863-1864.

Canadian Architect and Builder, 1888-1890.

Canadian Journal of Science, Literature and History [title varies], 1852-1878.

Canadian Manufacturer, 1882-1890.

Canadian Mining Review, 1882-1890.

Canadian Naturalist and Quarterly Journal of Science, 1856-1883.

Canadian Patent Office Record [title varies], 1873-1890.

The Canadian Record of Science, 1885-1890.

Colliery Engineer (Pennsylvania), 1881-1890.

Colliery Guardian, and Journal of the Coal and Iron Trades (London), 1858/59 to 1890.

Engineer (London), 1856-1866.

Engineering, An Illustrated Weekly Journal (London), 1866-1890.

Engineering Index [title varies] (United States), 1884-1900.

Engineering News-Record (Chicago-N.Y.), 1874-1890.

Geological Magazine (London), 1864-1890.

Inventor's Advocate and Journal of Industry: A Weekly British and Foreign Miscellany of Science, Inventions, Manufacturers and Arts, 1839-1841.

Journal of the American Statistical Association, 1888/89-1890.

Journal of the Franklin Institute, 1841-1845, 1867, 1877.

Journal of the Geological Society of London, 1846-1890.

Journal of the Royal Statistical Society (London), 1838-1890.

Mechanics' Magazine, Museum Register, Journal, and Gazette (London), 1840-1858.

Mineralogical Magazine (Mineralogical Society of London), 1877-1890.

The Mining Almanac (London), 1850, 1851.

Mining and Smelting Magazine (London), 1862-1865.

Mining Engineer (Institute of Mining Engineers, Newcastle), 1889-1890.

Mining Journal (London), 1860-1890 [incomplete].

Mining Magazine and Journal of Geology, Mineralogy, Chemistry, etc. (N.Y.), 1853-1858.

Mining World and Engineering Record (London), 1871-1890 [incomplete].

National Geographic Magazine (Washington), 1888-1890.

Popular Science (N.Y.), 1872-1890.
Proceedings and Transactions of the Royal Society of Canada, 1882-1890.
Proceedings of the American Society of Civil Engineers, 1873-1890.
Proceedings of the Institution of Mechanical Engineers (London), 1847-1890.
Proceedings of the Mining Institute of Cornwall, 1877-1884.
Proceedings of the Royal Geological Society (London), 1873-1880.
Repertory of Patent Inventions and Other Discoveries and Improvements in Arts, Manufacturing, and Agriculture (London), 1794-1828.
Reports and Proceedings of the Miners' Association of Cornwall and Devon, 1867-1890.
The School of Mines Quarterly (N.Y.), 1870-1890.
Scientific American (N.Y.), 1845-1888 [incomplete].
Scientific American Supplement, 1888-1890.
Thunder Bay Sentinel, 1875-1890.
Transactions of the American Institute of Mining Engineers, 1871-1890.
Transactions of the American Society of Civil Engineers, 1867-1890.
Transactions of the American Society of Mechanical Engineers, 1880-1890.
Transactions of the Engineering Institute of Canada, 1887-1890.
Transactions of the Institute of Mining Engineers (London), 1889-1890.
Transactions of the Manchester Geological and Mining Society, 1841-1890.
Transactions of the Mining Association and Institute of Cornwall, 1886-1893.
Transactions of the Mining Institute of Scotland, 1884-1890.
Transactions of the North-of-England Institute of Mining Engineers, 1852-1890.

E. Articles

Adams, J.E. "The Treatment of Gold and Silver Ores by Wet Crushing and Pan Amalgamation without Roasting." *Transactions of the American Institute of Mining Engineers* 2 (1873-74):159-71.
Allen, Robert C. "Collective Invention." *Journal of Economic Behavior and Organization* 4 (1983):1-23.
"American Diamond Drill Co." *Scientific American* 25, 13 (1871):200, 206.
"American Industries No. 63. The Manufacture of Power Drills for Mining, Excavating, Ec." *Scientific American* 43, 26 (1880):399, 402.
Ankli, Robert E.; Helsberg, H. Dan.; and Thompson, John Herd. "The Adoption of the Gasoline Tractor in Western Canada." In *Canadian Papers in Rural History,* edited by Donald H. Akenson, vol. 2, pp. 7-39. Gananoque: Langdale Press, 1980.
Argall, Philip. "Continuous Jigging Machinery." *Proceedings of the Mining Institute of Cornwall* 1, 9 (1884):337-51, 353-55.
Arthur, Elizabeth. "The Founding Father [Peter McKellar]." *Thunder Bay Historical Museum Society, Papers and Records* [*TBHMS Papers and Records*] 11 (1983):10-22.
Aschman, Homer. "The Natural History of a Mine." *Economic Geography* 46 (1970):172-89.
Attwood, Melville. "On the Milling of Gold Quartz: Amalgamation." *Scientific Canadian Mechanics' Magazine and Patent Office Record* 9 (October 1881): 291-95.
"Ball's Crushers." *Mechanics' Magazine* 6, 2 (February 1856): 193-94.

Barnett, J.D. "A Partial Bibliography of Petroleum." *Transactions of the Canadian Institute of Civil Engineers,* Pt. 1, 2 (1887):45-47.

Bassett, A. "The Diamond Drill." *Transactions of the North-of-England Institute of Mining Engineers* 23 (1873-74):179-96.

Beaumont, Major. "On Rock Boring by the Diamond Drill and Recent Application of the Process." *Proceedings of the Institution of Mechanical Engineers* (April 1875):82-125.

Bell, Robert. "The Petroleum Fields of Ontario." *Proceedings and Transactions of the Royal Society of Canada* 5 (1888):102-13.

Bennett, E.J. "Modern Stamping and Concentration Appliances." *Transactions of the Mining Association and Institute of Cornwall* 2, 2 (1889):56-66.

Bladen, M.L. "Construction of Railways in Canada to the Year 1885" (Part I). *Contributions to Canadian Economics* 1, 5 (1932):43-60.

Blackwell, S.H. "On Kind's Improved System of Boring." *Proceedings of the Institution of Mechanical Engineers* (1854):87.

Blainey, Geoffrey. "A Theory of Mineral Discovery: Australia in the Nineteenth Century," *Economic History Review* 2d ser., 23, 2 (1970):298-313.

Blake, Theodore A. "The Blake System of Fine Crushing." *Transactions of the American Institute of Mining Engineers* 13 (1884-85):232-48.

Blandy, John F. "Stamp Mills of Lake Superior." *Transactions of the American Institute of Mining Engineers* 2 (1873-74):208-15.

Bowler, Peter J. "The Early Development of Scientific Societies in Canada." In *The Pursuit of Knowledge in the Early American Republic: American Learned Societies from Colonial Times to the Civil War,* edited by A. Oleson and S.C. Brown, pp. 325-39. Baltimore: Johns Hopkins University Press, 1976.

Brain, Frank. "Electrical Pumping in Collieries." *Colliery Engineer* 8, 6 (1888):126-27.

Bruton, C. "Design of a Wind Hammer for Boring Rocks." *Transactions of the Royal Cornwall Polytechnic Institute* (1844):[n.p.].

"Bruton's Ore Dressing Frame." *Scientific American* 2, 37 (1847):289.

Bumstead, D.L. "Copper Smelting in Canada." *CIM Bulletin* 75, 846 (1982):36-40.

Burland, Jeffrey H. "The Metallurgy of Copper." *Scientific Canadian Mechanics' Magazine and Patent Office Record* 10 (March 1882):65-69.

"Canadian Railways – No. 29: Cobourg, Peterborough and Marmora Railway." *Engineering* 29 (January 1880):21-25.

Careless, J.M.S. "Frontierism, Metrpolitanism, and Canadian History." *Canadian Historical Review* 35 (1954):1-21.

Chandler, Alfred D., Jr. "Anthracite Coal and the Beginnings of the Industrial Revolution in the United States." *Business History Review* 46 (Summer 1972):143-81.

"Chapin's Improved Salt Evaporator." *Scientific American* 7, 7 (1862):97.

Chapman, E.J. "On the Wallbridge Hematite Mine, as Illustrating the Stock-Formed Mode of Occurrence of Certain Ore Deposits." *Proceedings and Transactions of the Royal Society of Canada* 4 (1885):23-26.

Christy, Samuel B. "The Growth of American Mining Schools and Their Relation to the Mining Industry." *Transactions of the American Institute of Mining Engineers* 23 (1893):444-65.

Cogging, Frederick G. "Notes on the Steam Stamp." *Engineering* 40 (29 January 1886):119; (5 February 1886):131-31; (6 February 1886):201-3.

Cornet, M.F.L. "On the Application of Machines Worked by Compressed-Air in the Collieries . . . in Belgium [and England]." *Transactions of the North-of-England Institute of Mining Engineers* 21 (1871-72):199-220.

Courtis, W.M. "The Wyandotte Silver Smelting and Refining Works." *Transactions of the American Institute of Mining Engineers* 2 (1873-74):89-100.

———. "The North Shore of Lake Superior as a Mineral-Bearing District." *Transactions of the American Institute of Mining Engineers* 5 (1876-77):473-87.

Cronin, Fergus. "North America's Father of Oil." *Imperial Oil Review* 39, 2 (1955):17-20.

———. "They Called It London-in-the-Bush." *Imperial Oil Review* 39, 4 (1955):21-24.

Cuthbertson, G.A. "Bruce Mines: A Brief History of Canada's Pioneer Copper Mine." *Canadian Mining Journal* 50 (July, 1939):424-26.

Daniel, William. "On Compressed-Air Machinery for Underground Haulage." *Proceedings of the Institution of Mechanical Engineers* (August 1874):204-33.

de Bresson, Christian. "Have Canadians Failed to Innovate? The Brown Thesis Revisited." *HSTC Bulletin* 6, 1 (1982):10-32.

DeGolyer, Everette Lee. "Seventy-Five Years of Progress in Petroleum." in *Seventy-Five Years of Progress in the Mineral Industry 1871-1946,* edited by A.B. Parsons, pp. 270-302. New York: AIME, 1947.

Degy, Julien. "The Kind-Chaudron Process for Sinking and Tubing Mine Shafts." *Transactions of the American Institute of Mining Engineers* 5 (1876-77):117-31.

Denis, Keith. "Oliver Daunais, the 'Silver King.' " *Thunder Bay Historical Museum Society Papers and Records* 2 (1974):12-21.

Douglas, James. "The Native Copper Mines of Lake Superior." *Canadian Naturalist* 7 (1875):318-36.

"Economical Method for Manufacturing Salt." *Scientific American* 11, 17 (1864):257.

Egleson, T. "Copper Mining on Lake Superior." *Transactions of the American Institute of Mining Engineers* 6 (1877-78):275-312.

Elford, Jean; and Phelps, Edward. "Oil, Then and Now." *Canadian Geographic Journal* 77 (November, 1968):164-71.

Ellis, W.H. "Nitro-Glycerine: Its History, Manufacture, and Industrial Application." *Canadian Journal* n.s. 14, 4 (1875):356-66.

Faucher, Albert, "The Decline of Shipbuilding at Quebec in the Nineteenth Century." *Canadian Journal of Economics and Political Science* 23, 2 (1957):195-215.

Ferguson, Eugene S. "Expositions of Technology, 1851-1900." In *Technology in Western Civilization,* edited by Melvin Kranzberg and Carroll W. Pursell, Jr., vol. 1, pp. 706-25. New York: Oxford University Press, 1967.

Ferguson, Henry T. "On the Mechanical Appliances Used for Dressing Tin and Copper Ores in Cornwall." *Proceedings of the Institution of Mechanical Engineers* Pt. 1 (July 1873):119-52.

Firestone, O.J. "Development of Canada's Economy, 1850-1900." In *Trends in the American Economy in the Nineteenth Century,* pp. 217-52. National Bureau of Economic Research, Studies in Income and Wealth. Princeton: Princeton University Press, 1960.

———. "Innovations and Economic Development – The Canadian Case." *Review of Income and Wealth* ser. 18, 8 (1972):399-419.

Fleming, Sir Sanford. "Notes on the Present Condition of the Oil Wells at Enniskillen."

Canadian Journal n.s. 46 (1863):246-49.

Frenchville, R.J. "The Results Obtained by the Cornish System of Dressing Tin Ores." *Transactions of the Mining Association and Institute of Cornwall* 1, 2 (1886):93-104.

"Frue Vanning Machine and Concentrator." *Mining World and Engineering Record* (11 March 1876).

"Garrison's Improved Salt Block." *Scientific American* 7, 13 (1862):193.

Gillis, Maurice. "He Took the 'Skunk' out of Oil." *Imperial Oil Review* 40, 3 (1956):8-10.

Graton, L.C. "Seventy-Five Years of Progress in Mining Geology." In *Seventy-Five Years of Progress in the Mineral Industry 1871-1946,* edited by A.B. Parsons, pp. 1-39. New York: AIME, 1947.

Griffin, Selwyn P. "Petrolia, Cradle of Oil-Drillers." *Imperial Oil Review* 14, 4 (1930):19-24.

Gwillum, J.C. "The Status of the Mining Profession." *Journal of the Canadian Mining Institute* 10 (1907):321-39.

Haber, Samuel. "The Professions and Higher Education in America: A Historical View." In *Higher Education and the Labor Market,* edited by Margaret S. Gordon, pp. 237-80. New York: McGraw-Hill, 1974.

Harley, C.K. "On the Persistence of Old Techniques: The Case of North American Wooden Shipbuilding." *Journal of Economic History* 33, 2 (1973):372-98.

Heinrich, O.J. "The Manhattan Salt Mine at Goderich, Canada." *Transactions of the American Institute of Mining Engineers* 6 (1877-78):125-44.

———. "The Industrial School for Miners at Drifton, Luzerne County, Pa." *Transactions of the American Institute of Mining Engineers* 9 (1880):390-95.

Henry, Louis. "The Training of a Mining Engineer." *Canadian Mining Review* 14, 2 (1895):220-21.

"Herman Frasch." *Dictionary of American Biography,* vol. 6, pp. 602-3. New York: Scribner's, 1931.

———. *National Cyclopedia of American Biography,* vol. 19, pp. 347-48. New York: James T. White, 1926.

"High Explosives: Dynamite or Giant Powder." *Scientific Canadian Mechanics' Magazine and Patent Record* 9 (August 1881):246-47.

Hodges, A.D., Jr. "Amalgamation at the Comstock Lode, Nevada: A Historical Sketch of Milling Operations at Washoe, and an Account of the Treatment of Tailings at the Lyon Mill, Dayton." *Transactions of the American Institute of Mining Engineers* 19 (1890-91):195-231.

"The Hoosac Tunnel." *Journal of The Franklin Institute* 3d ser., 44 (July 1867):9-10.

"The Hoosac Tunnel." *Scientific American* 22, 7 (1870):106.

Hunt, T.S. "The Goderich Salt Region." *Transactions of the American Institute of Mining Engineers* 5 (1876-77):538-60.

———. "The Apatite Deposits of Canada." *Transactions of the American Institute of Mining Engineers* 12 (1884):459.

Innis, H.A. "An Introduction to the Economic History of Ontario from Outpost to Empire." Ontario Historical Society, *Papers and Records* 30 (1934):111-23.

Jackson, Charles F. "Metal Mining Practice over Sixty Years." *Canadian Mining Journal* 60, 11 (1939):673-80.

Jernegan, J.L., Jr. "The Swansea Silver Smelting and Refining Works of Chicago." *Transactions of the American Institute of Mining Engineers* 4 (1875-76):35-53.

Johnson, Arthur M. "Expansion of the Petroleum and Chemical Industries, 1880-1900." In *Technology in Western Civilization,* edited by Melvin Kranzberg and Carroll W. Pursell, Jr., vol. 1, pp. 675-77. New York: Oxford University Press, 1967.

Jordon, Thomas. "On Rock-Drilling Machinery." *Proceedings of the Institution of Mechanical Engineers* (April 1874):77-100.

Kelly, Kenneth. "The Transfer of British Ideas on Improved Farming in Ontario during the First Half of the Nineteenth Century." *Ontario History* 63 (1971):103-11.

King, C.D. "Seventy-Five Years of Progress in Iron and Steel, Coke, Pig Iron and Ingot Manufacture." In *Seventy-Five Years of Progress in the Mineral Industry 1871-1946,* edited by A.B. Parsons, pp. 162-98. New York: AIME, 1947.

Krom, S.R. "Improvements in Ore-Crushing Machinery." *Transactions of the American Institute of Mining Engineers* 14 (1885-86):497-508.

Kuznets, Simon. "Inventive Activity: Problems of Definition and Measurement." In *The Rate and Direction of Inventive Activity: Economic and Social Factors.* Report of National Bureau of Economic Research, No. 13, pp. 19-51. Princeton: Princeton University Press, 1962.

Laist, Frederick. "Seventy-Five Years of Progress in Smelting and Leaching of Ores." In *Seventy-Five Years of Progress in the Mineral Industry,* edited by A.B. Parsons, pp. 126-61. New York: AIME, 1947.

"Lake Superior Native Copper Company's Mines." *Canadian Mining Review* 2, 4 (April 1884):3-4.

Lankton, Larry D. "The Machine *Under* the Garden: Rock Drills Arrive at the Lake Superior Copper Mines, 1868-1883." *Technology and Culture* 24, 1 (1983):1-37.

Layton, Edwin T., Jr. "Mirror Image Twins: The Communities of Science and Technology in 19th Century America." *Technology and Culture* 12, 4 (1971):562-80.

"Leschot's Diamond-Pointed Steel Drill." *Scientific American* 22, 18 (1870):282-83.

Locke, J.M. "The Brückner Revolving Furnace." *Transactions of the American Institute of Mining Engineers* 2 (1874):295-99.

Low, George. "Description of a Rock Boring Machine." *Proceedings of the Institution of Mechanical Engineers* (April 1865):179-200.

Luxmore, S. "Early Applications of Electricity to Coal Mining." *Proceedings of the Institute of Electrical Engineers* 126 (September 1979):869-74.

Macfarlane, Thomas. "Silver Islet." *Transactions of the American Institute of Mining Engineers* 8 (1879-80):226-53.

Maclean, John, "The Town That Rocked the Oil Cradle." *Imperial Oil Review* 39, 3 (1955):6-10.

Mansfield, Edward. "Technical Change and the Rate of Imitation." *Econometrica* (1961): 741-66.

Martell, B. "On the Carriage of Petroleum in Bulk on Over-Sea Voyages." *Engineering* 42 (30 July 1887):107-11.

"The Marquette Iron Region." *The School of Mines Quarterly* 3, 1 (1882): 35-40; 3, 2 (1882):103-17; 3, 3 (1882): 197-207; 3, 4 (1882):243-53.

Mather, Frank. "The Salt Deposits of New York." *Popular Science* 25 (1884):530-37.

McMillan, J.G. "Early Development of Ontario Mining." *Canadian Mining Journal* 54 (September 1943):561-64.

Michael, Rita. "Ironworking in Upper Canada: Charles Hayes and the Marmora Works."

CIM Bulletin 76, 849 (1983):132-35.

Miller, E. Willett. "Mineral Regionalism of the Canadian Shield." *Canadian Geographer* 13 (1959):17-30.

Miller, W.G. "Mines and Mining." In *Canada and Its Provinces,* edited by A. Shortt and A. Doughty, vol. 18, *Ontario,* pt. 2, pp. 613-45. Edinburgh: Constable, 1914.

Multauf, Robert P. "Mine Pumping in Agricola's Time and Later." Paper 7, USM Bulletin 218: *Contributions from the Museum of History and Technology* (1959):114-20.

Munroe, H.S. "The Losses in Copper Dressing at Lake Superior." *Transactions of the American Institute of Mining Engineers* 8 (1879-80):409-51.

———. "List of Books on Mining." *School of Mines Quarterly* 10, 2 (1889):176-84.

———. "The New Dressing Works of the St. Joseph Lead Company, at Bonne Terre, Missouri." *Canadian Mining Review* 9, 9 (September 1890):127-30.

Murray, Florence B. "Agricultural Settlement on the Canadian Shield." In *Profiles of a Province,* edited by Edith G. Firth, pp. 178-86. Toronto: Ontario Historical Society, 1967.

Nelson, Richard R. "Introduction." In *The Rate and Direction of Inventive Activity.* Report of the National Bureau of Economic Research, No. 13, pp. 3-16. Princeton: Princeton University Press, 1962.

Newell, Dianne. "Canada at the World's Fairs, 1851-1876." *Canadian Collector* 11, 4 (1976): 11-15.

———. "Published Government Documents as a Source for Interdisciplinary History: A Canadian Case Study." *Government Publications Review* 8A (1981):381-93.

———. "Technological Innovation and Persistence in the Ontario Oilfields: Some Evidence from Industrial Archaeology." *World Archaeology* 15, 2 (1983):184-95.

———. "'All in a Day's Work': Local Invention on the Ontario Mining Frontier." *Technology and Culture* (forthcoming).

Noble, David F. "Social Choice in Machine Design: The Case of Automatically Controlled Machine Tools and a Challenge for Labor." *Politics and Society* 8, 3-4 (1978):313-47.

North, Douglass C. "Location Theory and Regional Economic Growth." *Journal of Political Economy* 63 (June 1955): 243-58.

"Ore Testing Works and Recent Additions to the Assay Department of the School of Mines." *The School of Mines Quarterly* 6, 3 (1885):224-35.

Parsons, A.B. "History of the Institute [A.I.M.E.]." In *Seventy-Five Years of Progress in the Mineral Industry 1871-1946,* edited by A.B. Parsons, pp. 403-92. New York: AIME, 1947.

Parsons, C.S. "Sixty Years of Development in Ore Dressing." *Canadian Mining Journal* 60 (November 1939):693-711.

Peters, E.D., Jr. "The Sudbury Ore Deposits." *Transactions of the American Institute of Mining Engineers* 18 (1889-90):278-89.

———. "Sudbury Mines and Works." *Canadian Mining Review* 8, 9 (1889):123-36; 8, 10 (1889):135.

"Petroleum and Its Products." *Scientific American* Supplement (18 May 1872):335-42.

"Petroleum at Baku." *Scientific American* 40, 11 (1884):166; 40, 22 (1884):342-43.

Phelps, Edward. "The Canadian Oil Association — An Early Business Combination." *Western Ontario Historical Notes* 19 (September 1963):31-39.

———. "Foundations of the Canadian Oil Industry, 1850-1866." In *Profiles of a Province,* edited by Edith G. Firth, pp. 156-65. Toronto: Ontario Historical Society, 1967.

Pomfret, Richard. "Mechanization of Reaping in Nineteenth Century Ontario: A Case Study of the Pace and Causes of the Diffusion of Embodied Technical Change." *Journal of Economic History* 36 (June 1976):399-415.

"The Port Arthur Mines." *Canadian Mining Review* 8, 1 (January 1889):11-12.

Potter, William B. "Some Thoughts Relating to the American Institute of Mining Engineers and Its Mission [Presidential Address]." *Transactions of the American Institute of Mining Engineers* 17 (1888-89):485-94.

Pred, Allan R. "Industrialism, Initial Advantage, and American Metropolitan Growth." In *Geographic Perspectives on America's Past: Readings on the Historical Geography of the United States,* edited by David Ward, pp. 275-90. New York: Oxford University Press, 1979.

Rae, John B. "The Invention of Invention." In *Technology in Western Civilization,* edited by Melvin Kranzberg and Carroll W. Pursell, Jr., vol. 1, pp. 327-36. New York: Oxford University Press, 1967.

Rand, A.C. "A New Rock Drill without Air Cushion." *Transactions of the American Institute of Mining Engineers* 13 (1884-85):249-53.

"Rand Rock Drills." *Colliery Engineer* 9, 12 (July 1889):270.

Rathbone, Edgar P. "On Copper Mining in the Lake Superior District." *Proceedings of the Institution of Mechanical Engineers* (February 1887): 86-123.

Reingold, Nathan. "U.S. Patent Office Records as Sources for the History of Invention and Technological Property." *Technology and Culture* 1, 1 (1960):156-67.

Richards, J.H. "Lands and Policies: Attitudes and Controls in the Alienation of Lands in Ontario during the First Century of Settlement." *Ontario History* 50 (1958): 192-209.

———. "Population and the Economic Base in Northern Hastings County, Ontario." *Canadian Geographer* 11 (1958): 22-31.

Richards, R.H. "American Mining Schools." *Transactions of the American Institute of Mining Engineers* 15 (1886-87):309-40; 809-19.

Richardson, John. "On the Application of Portable Engines for Mining Purposes." *Proceedings of the Institution of Mechanical Engineers* (July 1873):167-201.

Rivolt, E. "Visit to the Lake Superior Region, 1854." *Mining Magazine* 2, 6 (1854):28-37; 97-106; 207-16.

Robb, Charles. "On the Petroleum Springs of Western Canada." *Canadian Journal* n.s. 6 (1861):313-23.

Roebling, John A. "American Manufacture of Wire Ropes for Inclined Planes, Standing Riggings, Mines, Tillers, Ec." *Journal of the Franklin Institute* 3d ser., 7 (Jan., 1844):57-60.

Rolker, Charles M. "The Allouez Mine and Ore Dressing as is Practiced in the Lake Superior Copper District." *Transactions of the American Institute of Mining Engineers* 5 (1876-77):584-611.

Rosenberg, Nathan. "Technological Change in the Machine Tool Industry, 1840-1910." *Journal of Economic History* 23 (1963):414-43.

———. "The Direction of Technological Change: Inducement Mechanisms and Focusing Devices." *Economic and Cultural Change* 18 (1969):1-24.

———. "Factors Affecting the Diffusion of Technology." *Explorations in Economic History* 10 (January 1973):3-33.

———. "American Technology: Imported or Indigenous?" *American Economic Review* 67 (1977):21-31.

Rothwell, R.P. "The Gold-Bearing Mispickel Veins of Marmora, Ontario, Canada." *Transactions of the American Institute of Mining Engineers* 9 (1881):409-20.

Rouse, A.K. "Borlase's Cornwall." *History Today* 23, 12 (1973):879-83.

Russell, Loris. "Abraham Gesner." *Dictionary of Canadian Biography*, vol. 9, pp. 308-12. Toronto: University of Toronto Press, 1977.

———. "Invention and Discovery." In *Science, Technology and Canadian History*, edited by R.A. Jarrell and N.R. Ball, pp. 120-28. Waterloo: Wilfrid Laurier University Press, 1980.

"The Salt Industry of Onondaga County, N.Y." *Scientific American* 63, 24 (1890):367, 373.

Samuel, Raphael. "The Workshop of the World: Steam Power and Hand Technology in Mid-Victorian Britain." *History Workshop Journal* 3 (1977):6-72.

Saywell, John T. "The Early History of Canadian Oil Companies: A Chapter in Canadian Business History." *Ontario History* 53 (1961):67-72.

Schmookler, Jacob. "The Interpretation of Patent Statistics." *Journal of the Patent Office Society* 32, 2 (1950):123-46.

———. "The Utility of Patent Statistics." *Journal of the Patent Office Society* 35, 6 (1953): 407-12.

———. "Patent Office Statistics as an Index of Inventive Activity." *Journal of the Patent Office Society* 35, 8 (1953):539-50.

Sclanders, Ian. "He Gave the World a Brighter Light." *Imperial Oil Review* 39, 1 (1955): 23-25.

———. "The Amazing Jake Englehart." *Imperial Oil Review* 39, 4 (1955):2-7.

Scott, Benjamin S. "Oil Refining in London." *Western Ontario Historical Notes* 6, 3/4 (1948):38-45.

Scott, Beryl H. "The Story of Silver Islet." *Ontario History* 49 (1957):125-37.

Seely, Bruce E. "Blast Furnace Technology in the Mid-19th Century: A Case Study of the Adirondack Iron and Steel Company." *IA* 7, 1 (1981):27-54.

Shallenberg, Richard H. "Evolution, Adaptation and Survival: The Very Slow Death of the American Charcoal Iron Industry." *Annals of Science* 32, 4 (July 1975):341-58.

Small, George W. "Notes on the Stamp Mill and Chlorination Works of the Plymouth Consolidated Gold Mining Company, Amadore County, California." *Transactions of the American Institute of Mining Engineers* 15 (1886-87):305-8.

Smith, Cyril Stanley. "Mining and Metallurgical Production." In *Technology in Western Civilization*, edited by Melvin Kransberg and Carroll W. Pursell, Jr., vol. 1, pp. 143-46. New York: Oxford University Press, 1967.

Solo, Carolyn Shaw. "Innovation in the Capitalist Process: A Critique of the Schumpeterian Theory." *Quarterly Journal of Economics* 65, 3 (1951):417-28.

Spaulding, H.C. "Electric Power-Transmission in Mining Operations." *Transactions of the American Institute of Mining Engineers* 19 (1890-91):258-88.

Spillsbury, E. Gybbon. "Rock-Drilling Machinery." *Transactions of the American Institute of Mining Engineers* 3 (1874-75):144-50.

Spragge, George W. "Colonization Roads in Canada West, 1850-1867." *Ontario History* 49 (1957):1-17.

Stelter, Gilbert A. "The Origins of a Company Town: Sudbury in the Nineteenth Century." *Laurentian University Review* 3, 3 (1971):1-35.

———. "Community Development in Toronto's Commercial Empire: The Industrial Towns of the Nickel Belt, 1883-1931." *Laurentian University Review* 6, 3 (1974): 3-53.

Stevens, William. "The Prospects of the Lake Superior Mining Region." *Mining Magazine* 2, 2 (1854):149-53.

Taggart, Arthur F. "Seventy-Five Years of Progress in Ore-Dressing." In *Seventy-Five Years of Progress in the Mineral Industry 1871-1946,* edited by A.B. Parsons, pp. 82-125. New York: AIME, 1947.

Talbot, Allan G. "Ontario Origin of the Canadian Explosives Industry." *Ontario History* 56 (1964): 37-44.

Tann, Jennifer. "Fuel-Saving in the Process Industries during the Industrial Revolution: A Study in Technological Diffusion." *Business History* 15 (July 1975):149-59.

Temin, Peter. "Steam and Waterpower in the Early Nineteenth Century." *Journal of Economic History* 26 (1966):187-205.

Thompson, Wilbur R. "Locational Differences in Inventive Effort and Their Determinants." In *The Rate and Direction of Inventive Activity.* Report of the National Bureau of Economic Research, No. 13, pp. 253-67. Princeton: Princeton University Press, 1962.

Twite, Charles. "Technical Education of Miners." *Proceedings of the Mining Institute of Cornwall* 1, 7 (1883):201-14.

Uselding, Paul; and Juba, Bruce. "Biased Technical Progress in American Manufacturing." *Explorations in Economic History* 11 (1973):55-72.

Vogel, Robert M. "Tunnel Engineering: A Museum Treatment," Paper 41, USM Bulletin 240. *Contributions from the Museum of History and Technology* (1964):201-40.

———. "Building in the Age of Steam." In *Building Early America. Contributions toward the History of a Great Industry,* edited by Charles E. Petersen, pp. 119-34. Radnor, Pa.: Chilton Book Co., 1976.

Wadsworth, M.E. "Some Statistics on Engineering Education." *Transactions of the American Institute of Mining Engineers* 27 (1897):712-31.

Walker, James. "Blasting Explosives." *Transactions of the Mining Institute of Scotland* 6 (1884-85):94-108.

Watkins, Melville. "Technology in Our Past and Present." In *Canada: A Guide to the Peaceable Kingdom,* edited by William M. Kilbourne, pp. 285-92. Toronto: MacMillan, 1970.

"William Plummer." *Canadian Mining Review* 9, 6 (1890):82-83.

Williams, H.J. Carnegie. "The Bruce Mines, Ontario, 1846-1906." *Journal of the Canadian Mining Institute* 10 (1907):149-69.

Williamson, Harold; and Andreano, Ralph. "Competitive Structure of the American Petroleum Industry 1880-1911." In *Oil's First Century.* Centennial Seminar on the History of the Petroleum Industry. Cambridge, Mass.: Harvard University Press, 1959.

Wood, De Volson. "Drilling Machines at the Hoosac Tunnel." *Journal of the Franklin*

Institute 3d ser., 44 (August 1867):83-86.
"Woodruff's Plan for Lining Oil Barrels." *Scientific American* 13, 26 (1865):399.
Zaslow, Morris. "The Frontier Hypothesis in Recent Historiography." *Canadian Historical Review* 29 (1948):153-66.
———. "The Ontario Boundary Question." In *Profiles of a Province,* edited by Edith G. Firth, pp. 107-17. Toronto: Ontario Historical Society, 1967.
———. "The Yukon: Northern Development in a Canadian-American Context." In *Regionalism in the Canadian Community, 1867-1967,* edited by Mason Wade, pp. 180-97. Toronto: University of Toronto Press, 1969.
———. "Edward Barnes Borron, 1820-1915, Northern Pioneer and Public Servant Extraordinary." In *Aspects of Nineteenth-Century Ontario,* edited by Frederick H. Armstrong, et al. pp. 297-311. Toronto: University of Toronto Press, 1974.

UNPUBLISHED WORKS

A. Manuscript Collections

Montreal. McGill University Archives. J.W. Dawson Papers.
———. ———. Registers of Matriculation.
———. ———. William Edmond Logan Papers.
———. McGill University Rare Book Collection. Thomas Macfarlane Manuscript Collection.
Ottawa. Department of Consumer and Corporate Affairs. Patent Files.
———. Public Archives of Canada. Allan Macdonell Mining Papers [Quebec and Lake Superior Mining Company].
———. ———. British North American Mining Company Papers.
———. ———. Department of Mines. Mineral Statistics.
———. ———. Engineering Institute of Canada Collection. Membership Files.
———. ———. Macdonell and Foster Papers [patent agents].
———. ———. Robert Bell Papers.
———. ———. T.A. Keefer Papers.
———. ———. Winnifred Philpot Papers [Silver Islet].
Toronto. Ontario Archives. A.N. Buell Papers.
———. ———. Canada Company Papers.
———. ———. Cross Family Papers [Silver Islet].
———. ———. Department of Crown Lands [Ontario].
———. ———. Gillies Brothers Lumber Company Records.
———. ———. John Fisken Papers [oil merchant].
———. ———. William Hamilton Merritt Papers.
Winnipeg. Public Archives of Manitoba. Hudson's Bay Company Archives.

B. Other.

Ball, Norman Roger. "Canadian Petroleum Technology during the 1860's." Master's thesis, University of Toronto, 1972.

Blair, Alexander Marshall. "Surface Extraction of Non-Metallic Minerals in Ontario Southwest of the Frontenac Axis." Ph.D. dissertation, University of Illinois, 1965.

Daum, Arnold R. "The Illumination Revolution and the Rise of the Petroleum Industry, 1850-1863." Ph.D. dissertation, Columbia University, 1957.

DeVecchi, Vittorio Maria Giuseppi. "Science and Government in Nineteenth-Century Canada." Ph.D. dissertation, University of Toronto, 1978.

Drell, Bernard. "The Role of the Goodman Manufacturing Company in the Mechanization of Coal Mining." Ph.D. dissertation, University of Chicago, 1930.

Evans, Margaret. "Oliver Mowat and Ontario, 1872-1896." Ph.D. dissertation, Harvard University, 1967.

Harkness, Robert B. "Makers of Oil History, 1850-1880." Unpublished manuscript, Regional History Collection, University of Western Ontario Library, n.d.

Hogarth, D.D. "Apatite in the Ottawa and Kingston Areas: A Flashback." Paper read before the symposium, Historic South Shield Mining Resources and Future Public Management, Kemptville, Ontario, 23 June 1976.

Hood, Robin. "The History of [Iron] Mining [at Marmora]." Unpublished manuscript prepared for the Crowe Valley Conservation Authority, Ontario, 1975[?].

Jestin, Warren James. "Provincial Policy and the Development of the Metallic Mining Industry in Northern Ontario: 1845-1920." Ph.D. dissertation, University of Toronto, 1977.

Johnson, Arthur Menzies. "The Development of American Petroleum Pipe Lines: A Study in Enterprise and Public Policy, 1862-1906." Ph.D. dissertation, Vanderbilt University, 1954.

Menanteau-Horta, Dario. "Diffusion and Adoption of Agricultural Practices among Chilean Farmers: A Sociological Study of the Processes of Communication and Acceptance of Innovations as Factors Related to Social Change and Agricultural Development in Chile." Ph.D. dissertation, University of Minnesota, 1967.

MacLeod, Donald. "Mines, Mining Men, and Mining Reformers: Changing the Technology of Nova Scotian Gold Mines and Colleries, 1858 to 1910." Ph.D. dissertation, University of Toronto, 1981.

Michael, Rita. "Report to the Ontario Heritage Foundation on the 1982 Archaeology of Marmora." Unpublished manuscript prepared for the Ontario Heritage Foundation, 1982.

Miller, Marilyn G. "Small Scale Mining in the South Shield Region of Eastern Ontario." Unpublished study prepared for the Ontario Ministry of Culture and Recreation and Ministry of Natural Resources, March, 1976.

Nakitsas, George. "The Employment Effects of Technique Choice: The Canadian Pulp and Paper Industry, 1951-1973." Master's thesis, McGill University, 1976.

Newell, Dianne. "Technological Change in a New and Developing Country: A Study of Mining Technology in Canada West-Ontario, 1841-1891." Ph.D. dissertation, University of Western Ontario, 1981.

———. "Professionalization of Mining Engineers in Late Nineteenth-Century Canada." Unpublished paper read before the Workshop on Professionalization in Modern Society, University of Western Ontario, 13-15 March 1981.

Page, Robert. "The Early History of Imperial Oil 1880-1900: Technology, Markets, Capital, and Continental Integration." Paper read before the Annual Meeting of the Canadian Historical Association, Vancouver, 6-8 June 1983.

Petersen, James Otto. "The Origins of Canadian Gold Mining: The Part Played by Labour in the Transition from Tool Production to Machine Production." Ph.D. Dissertation, University of Toronto, 1977.

Phelps, Edward. "John Henry Fairbank of Petrolia (1831-1914)." Master's thesis, University of Western Ontario, 1965.

Pomfret, Richard William Thomas. "The Introduction of the Mechanical Reaper in Canada, 1850-70: A Case Study in the Diffusion of Embodied Technical Change." Ph.D. dissertation, Simon Fraser University, 1974.

Potter, R.R. "The Woodstock Ironworks, Carleton County, New Brunswick." Paper read before the Annual Conference of Metallurgists. Ottawa, August, 1982.

Skeoch, Allan. "Technology and Change in Nineteenth-Century Ontario Agriculture." Master's thesis, University of Toronto, 1976.

Stacey, Duncan A. "Technological Change in the Fraser River Salmon Canning Industry, 1871-1912." Master's thesis, University of British Columbia, 1978.

Thoenon, Eugene David. "Petroleum Industry in West Virginia before 1900." Ph.D. dissertation, University of West Virginia, 1956.

Watson, Dennis McLean. "Frontier Movement and Economic Development in Northeastern Ontario, 1850-1914." Master's thesis, University of British Columbia, 1971.

Wright, Margaret Anna. "The Canadian Frontier, 1840-1867." Ph.D. dissertation, University of Toronto, 1943.

Wyman, Walker De Marquis, Jr. "The Underground Miner, 1860-1910: Labor and Industrial Change in the Northern Rockies." Ph.D. dissertation, University of Washington, 1971.

Zerker, Sally F. "Centralization in the International Oil Industry." Paper read before the Annual Meeting of the Canadian Historical Association. Vancouver, 6-8 June 1983.

NAME INDEX

SUBJECT INDEX